REVISE BTEC NATIONAL
Animal Management

REVISION GUIDE

Series Consultant: Harry Smith

Authors: Natalia Betts, Laura Johnston and Leila Oates

A note from the publisher

While the publishers have made every attempt to ensure that advice on the qualification and its assessment is accurate, the official specification and associated assessment guidance materials are the only authoritative source of information and should always be referred to for definitive guidance.

This qualification is reviewed on a regular basis and may be updated in the future. Any such updates that affect the content of this Revision Guide will be outlined at **www.pearsonfe.co.uk/BTECchanges**. The eBook version of this Revision Guide will also be updated to reflect the latest guidance as soon as possible.

Introduction

Which units should you revise?

This Revision Guide has been designed to support you in preparing for the externally assessed units of your course. Remember that you won't necessarily be studying all the units included here – it will depend on the qualification you are taking.

BTEC National Qualification	Externally assessed units
Extended Certificate	Unit 3 Animal Welfare and Ethics
For both: Foundation Diploma Diploma	Unit 2 Animal Biology Unit 3 Animal Welfare and Ethics
Extended Diploma	Unit 1 Animal Breeding and Genetics Unit 2 Animal Biology Unit 3 Animal Welfare and Ethics

Your Revision Guide

Each unit in this Revision Guide contains two types of pages, shown below.

Content pages help you revise the essential content you need to know for each unit.

Skills pages help you prepare for your exam or assessed task. Skills pages have a coloured edge and are shaded in the table of contents.

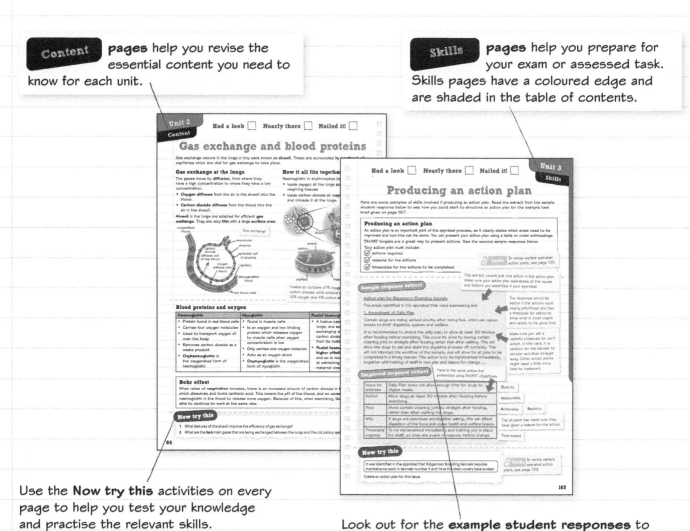

Use the **Now try this** activities on every page to help you test your knowledge and practise the relevant skills.

Look out for the **example student responses** to revision questions or tasks on the skills pages. Post-its will explain their strengths and weaknesses.

Contents

Workbook also available for externally assessed units ISBN 9781292149998

Unit 3: Animal Welfare and Ethics

A small bit of small print
Pearson publishes Sample Assessment Material and the
Specification on its website. This is the official content
and this book should be used in conjunction with it.
The questions in *Now try this* have been written to help
you test your knowledge and skills. Remember: the real
assessment may not look like this.

Reasons for breeding

The majority of animals bred are for food production. Others are bred for leisure, sport, the pet trade and in conservation programmes for endangered species.

Breeding for the pet trade

A range of animals are bred for the pet trade, including **companion animals** (such as dogs, cats and rabbits) and **exotics** (such as snakes, lizards and tortoises).

Working animals

Examples of working animals include:

☑ horses (for pulling carriages and ceremonial duties)

☑ support dogs (guide dogs, hearing dogs, diabetes and epilepsy dogs)

☑ rescue dogs (police, search and rescue)

☑ sniffer dogs (explosives, drugs, dead bodies).

Breeding for food production

Humans have **farmed** animals for thousands of years.

* **Extensive farming:** Some species, such as upland sheep, are bred in managed but natural, wild situations. This is where there is a large area with a small number of produce.

* **Intensive farming:** Poultry can be produced in battery situations, in a small area with a large number of produce.

Pig and cattle breeding tends to be carefully managed. Farmers may use reproductive technologies in some instances. Animals are also bred for their coats and skins, although much of this is a by-product of food production. Animal breeding can have a large impact on a country's economy, for example, successful breeding programmes for livestock can increase availability of food and high quality animals for trade.

The majority of animals bred are for food production. Chickens are bred for egg and meat production. Livestock can be bred specially to improve production. This can include not only selection for certain desirable characters or traits (for example, milk production), but also selection of livestock which show resistance to disease.

Sport animals

Many different species are used in **sports**, such as:

* greyhounds in racing
* horses in racing, eventing and trekking
* bulls in rodeos
* fishing
* gundogs.

A thoroughbred horse bred specifically for its speed.

Conservation breeding

Conservation breeding is used for maintaining and increasing the numbers of endangered animals. International exchanges help to increase **genetic diversity**, reducing the risk of disease. For example, due to conversation work, tiger numbers have increased from 40 animals in the wild in 1940s to around 500 in 2007.

Now try this

What are the main reasons for breeding animals?

Consider all the different reasons that people may keep animals.

Genetic terminology

The study of genetics relies on a vocabulary all of its own. Understanding the terms is central to being able to interpret and answer examination questions correctly.

Genes and alleles

All organisms of a species have the same genes. Patterns of inheritance are not about different genes, therefore, but are about inheritance of variants of the genes called **alleles**.

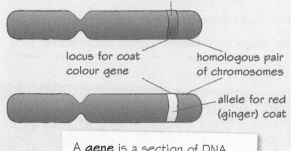

allele for black coat

locus for coat colour gene

homologous pair of chromosomes

allele for red (ginger) coat

A **gene** is a section of DNA that determines the structure of a polypeptide. An allele is a possible variant of the gene.

Defining genetic terms

- **Genotype** describes all of the genetic information in an organism, some of which is expressed to affect characteristics.

allele for black coat

- **Phenotype** describes the physical and biochemical characteristics of an organism. This will be determined by the genotype but will also be influenced by the environment.

A black coat is the phenotype produced by the black alleles for coat colour in the genotype.

Describing phenotypes

● dominant

- **Dominance** describes a phenotype that is expressed whenever one dominant allele coding for it is present, whichever other alleles are present.

○ recessive

- **Recessive** describes a phenotype that is only expressed when both alleles (recessive alleles) code for the phenotype.

●● black coat   ●◓ black coat   ○○ red (ginger) coat

'Gene' and 'allele' are often used as if they mean the same thing but they are very different. Make sure that you understand how. All cows have a gene for coat colour but they may have different versions, or alleles, of the gene.

Describing genotypes

Diploid organisms have two copies of their genes, one from each parent, which may or may not be the same allele.

- **Homozygous** describes a genotype in which both alleles for a gene are the same. A homozygous individual is a homozygote.

●●
homozygous for the dominant phenotype

○○
homozygous for the recessive phenotype

- **Heterozygous** describes a genotype where the two alleles for a gene are different. A heterozygous individual is a heterozygote.

●○
heterozygous

Multiple alleles

Some genes have more than two possible alleles in a population, for example, the human ABO blood groups are determined by a gene that has three alleles. It is important to remember, however, that a diploid individual can carry only two of the possible alleles.

Now try this

Explain why an animal needs to have two copies of a recessive allele to have a recessive phenotype.

Mendelian genetics and monohybrid crosses

Gregor Mendel's work on pea plants formed the basis of our understanding of genetics.

Mendel's three laws of inheritance

1 Law of **segregation** – during the formation of gametes, chromosome pairs are separated.

2 Law of **independent assortment** – alleles on pairs of chromosomes are allocated to gametes randomly.

3 Law of **dominance** – some alleles are dominant to others. In any organism, dominant alleles will be expressed over recessive alleles as the phenotype depends on the presence of dominant alleles.

Segregation – Mendel's first law

Mendel's law of **segregation** means that an individual can pass only one allele for a characteristic onto a gamete. In a heterozygote, there will be an equal probability of this being either of their two different alleles.

Dominance – Mendel's third law

Mendel's third law states that some alleles are dominant over others. This means that the feature the dominant allele codes for will always be expressed if it is present in the animal's genome. Dominant alleles are represented by capital letters. Recessive alleles are represented by lower case letters.

Monohybrid crosses

Monohybrid inheritance looks at a characteristic determined by a **single gene**.

Genetic diagrams can predict possible outcomes of a cross when constructed properly.

Show what you already know about the cross. Mixing up genes and alleles can lead you astray.

Remember that only one gamete can be passed to offspring by a parent. It wouldn't be wrong to write 'A or A' here instead of just A, but if you understand that this is a probability then you'll see that it isn't needed – all offspring will inherit A.

F1 is the first generation
F2 is the second generation

Always give the expected proportions or ratio for offspring. A **3:1 ratio** is a normal result from crossing two heterozygotes with normal dominance.

Example of a genetic diagram

Genes	Fur colour
Alleles	Brown fur (dominant) A
	White fur (recessive) a
Parental phenotype	homozygous brown fur x white fur
Parental genotype	AA × aa
Possible gametes	A a
F1 genotypes	

		a
		a
A	Aa

F1 phenotypes   All brown fur
F1 genotypes   Aa × Aa
Possible gametes   A or a   A or a
F2 genotype

	A	a
A	AA	Aa
a	Aa	aa

A Punnett square showing the possible offspring of the first generation (this is called the second generation).

F2 phenotype   ¾ brown fur
¼ white fur

Now try this

A reptile dealer is wanting to produce corn snakes which lack the black pigment (amelanistic), a recessive trait. The female chosen to breed has the genotype Bb.

What possible genotypes could the male corn snakes have to successfully produce amelanistic offspring? Explain your answer using Punnett squares.

3

Dihybrid crosses

Two non-interacting **unlinked genes** will be inherited **independently** of each other. The expected ratios of offspring with each characteristic will be the same as if the genes were considered separately.

Independent assortment

Mendel's second law states that the inheritance of one characteristic will have no effect on the inheritance of another. It has become clear that this only applies when genes are not linked on the same chromosome.

Dihybrid crosses look at the pattern of inheritance when two genes are considered at the same time.

Dihybrid genetic diagrams

Diagrams for dihybrid crosses should be set out similarly to those of monohybrid crosses (see page 3). This example uses the genes for coat colour (B = black, b = red) and the presence of horns (H = horns, h = no horns) in cattle.

The main difference is in showing the genotype for both genes at the same time:

	Monohybrid	Dihybrid
Parent genotypes	BB × bb HH × hh	BBHH × bbhh
Possible gametes	B b H h	BH bh
F₁ genotypes	Bb Hh	BbHh

The next generation shows all possible combinations of inheritance for the two genes:

F₁ genotypes	BbHh	BbHh
Possible gametes	BH Bh bH bh	BH Bh bH bh

F₂ genotypes

	BH	Bh	bH	bh
BH	BBHH	BBHh	BbHH	BbHh
Bh	BBHh	BBhh	BbHh	Bbhh
bH	BbHH	BbHh	bbHH	bbHh
bh	BbHh	Bbhh	bbHh	bbhh

This works out as a **9:3:3:1 ratio**:

9	3	3	1
black horned	black no horns	red horned	red no horns

Notice the pattern for each of the two genes still works out independently as the expected 3:1 ratio.

- For the colour gene, the 3:1 ratio is shown by repeated genotypes in each quarter of the grid.
- For the horns gene, the 3:1 ratio repeats in each quarter of the grid.

Explaining the 9:3:3:1 ratio

1 The four different types of gamete arise because the chromosomes arrange randomly during **meiosis**.

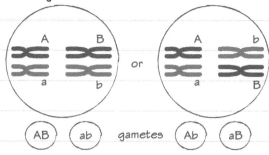

2 The ratio is all combinations of two 3:1 ratios:
(3B + 1b)(3H + 1h) = (9BH + 3Bh + 3bH + bh)

3 The probability of black **and** horned is given by the probability of black times the probability of horned:
$$\frac{3}{4} \times \frac{3}{4} = \frac{9}{16}$$

Calculating probability

Cattle have black and red alleles for coat colour and horned and no horn alleles. All of the offspring from a cross between horned black cattle and hornless red cattle were horned and black. Calculate the probability that a cow produced from a cross between one of these offspring and a hornless red cow will be horned.

Genotypes	BbHh	bbhh
Possible gametes	BH Bh bH bh	bh

Offspring

	BH	Bh	bH	bh
bh	BbHh	Bbhh	bbHh	bbhh

Phenotypes	black horned	black no horns	red horned	red no horns
Probability	$\frac{1}{4}$	$\frac{1}{4}$	$\frac{1}{4}$	$\frac{1}{4}$

Probability of horns is $\frac{1}{4} + \frac{1}{4} = \frac{1}{2}$ (0.5)

Now try this

In some breeds of sheep, polled (p) is recessive to horned (P) and white wool (W) is dominant to coloured (w). A polled ewe with white wool (WWpp) has been chosen to crossbreed with a horned ram with coloured wool (wwPp).

1 What possible gametes could be produced?

2 Calculate the phenotypic ratio of the offspring.

Gene interactions

Two or more genes on the same **chromosome**, or on different chromosomes, can **interact** with one another, affecting the phenotype produced. These interactions can affect how an animal looks, or have health implications.

Incomplete dominance – neither allele is completely dominant to the other. This produces a new phenotype where the heterozygous animal will look different to both homozygous animals.

Lethal alleles – these lead to death in the animals that have them. They can be recessive, dominant or show other interactions such as lethal white overo in horses.

Codominance – neither allele is completely dominant to the other. Both alleles are shown at the same time in the phenotype.

Gene interactions

Multiple alleles – this is where three or more alternative alleles of a single gene are possible (but only two are present in an individual organism).

Epistatic effects – this is where one gene affects the expression of another gene.

Sex-influenced – the phenotype depends on the gender of the animal even though the gene controlling that characteristic is not on the sex chromosome. It is influenced by sex hormones.

Sex-linkage – this is when genes are carried on the sex chromosomes (X or Y).

Incomplete dominance

The alleles for black and white feather colours interact to produce blue colouring in Andalusian fowl.

Andalusian chicken.

Codominance

Alleles for white and brown coat colour are both expressed in this interaction.

Cow with white and brown coat.

Now try this

Describe the effect of incomplete dominance.

Make sure you use scientific terminology such as allele and phenotype in your answer.

Sources of variation

Variation between individuals of one species is influenced by genes and the environment.

Variation

Existing **genetic variation** in an individual can be rearranged into different combinations by **random assortment** and **crossing over (recombination)** during **meiosis** when gametes are formed. Further variation arises due to the **random fertilisation** of gametes during sexual reproduction. The environment can also affect the expression of genes and the resulting phenotype.

 Links See page 3 for more information on Mendel's second and third laws.

Mutations

New genetic variation arises from changes in the DNA of a cell called **mutations**. They can be beneficial, harmful or neutral. Mutations can occur spontaneously, due to mistakes during **DNA replication**, or be the result of environmental factors (**mutagens**).

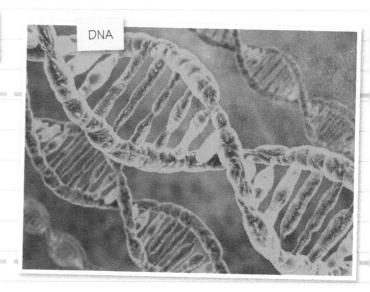
DNA

DNA

DNA is a double-stranded molecule made up of a sequence of pairs of four chemical bases: adenine (A), guanine (G), cytosine (C) and thymine (T). This code is read in sets of three pairs called **codons** which code for specific amino acids.

Types of mutation

Insertion and deletion mutations are also known as **frameshift mutations**, as they can alter the point from where a codon is read.

Mutation type	Description	
Point/substitution	One base pair is swapped for another, e.g. AAC|GTC|TCC to AAG|GTC|TCC. There are three types:	
	Silent	The change occurs in non-coding regions of DNA so it does not affect the phenotype.
	Missense	The codon now codes for a different amino acid.
	Nonsense	The change codes for a STOP codon, resulting in an incomplete protein.
Insertion	Extra base pairs are inserted into the DNA sequence, e.g. AAC|GTC|TCC to AAC|GAT|CTC|C	
Deletion	Base pairs are lost from the DNA, e.g. AAC|GTC|TCC to AAC|GTC|TC	
Duplication	A section of the DNA is repeated, e.g. AAC|GTC|TCC to AAC|GTC|TCC|TCC	
Translocation	Sections of DNA either swap places or chromosomes join together.	

Now try this

Explain why it is unlikely that two gametes will have the same genetic content despite the large numbers produced.

Selective breeding

Humans can choose which animals breed together, which is known as selective breeding.

Natural selection

Genetic variants for phenotypes (traits) that result in the animal being better able to survive and reproduce increase in frequency in the population. This is **natural selection**. Variations that enable an animal to better survive a changing environment are more likely to be passed on, and that trait becomes more common in the species. This is known as **adaptation**.

Selective breeding

Humans have been artificially selecting which traits become prevalent in groups of animals for thousands of years, to produce animals with desirable characteristics for particular purposes. This **non-random mating** leads to an overall change in the genotype of the population over many generations.

> **Links** Advances in animal breeding have been assisted by modern reproductive technologies (page 18) and genetic engineering techniques (page 33).

Positive assortive mating

Mating animals with **similar phenotypes** or features attempts to increase the occurrence of these traits in the offspring, for example, mating two corn snakes with wild type markings.

Negative assortive mating

Mating animals with **opposite phenotypes** or features can lead to more variety between offspring. For example, a wild type corn snake that has predominantly yellow and orange blotches can be mated with a striped anerythristic corn snake.

Positive assortive mating.

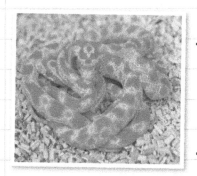

= Breeding two orange marked corn snakes will increase the chances of offspring continuing to have similar orange markings.

Negative assortive mating.

Anerythristic corn snakes do not have the red pigment.

= Breeding an anerythristic corn snake and an orange corn snake will increase the genetic diversity of the offspring.

Closely-related animals may be intentionally selected for breeding (**inbreeding**) to maintain the bloodline, for example, mating two dogs with the same parents or grandparents. This decreases the genetic diversity of the population, which can make it more likely that particular traits are inherited. However, it can result in **inbreeding depression**. This increases the chance of an inherited disease occurring, as homozygous mutations become more likely to be expressed.

Choosing to breed animals that are not closely related (outbreeding), or not related at all, can reduce the chance of inherited disease and increase the genetic diversity of the offspring. However, **outbreeding depression** can occur, where the offspring show the desired traits much less strongly. In the wild, this could result in the offspring not being suitably adapted to their environment.

Now try this

A beef cattle farmer is developing a breeding plan to cross Aberdeen Angus with Limousin.

Identify the type of mating and explain the implications it will have for future generations.

Desirable characteristics

Desirable characteristics are **traits** that breeders aim to show in the **phenotype** of offspring.

Mammals

The role of the animal is what determines their **desirable characteristics**. For example:

Pets

- markings
- sociability
- breed characteristics.

Livestock

- milk yield
- muscle mass
- body condition score
- docile nature
- fertility
- straightforward **parturition**
- animals that are acclimatised to certain environments, for example, hardiness in upland sheep
- disease resistance.

> Parturition is giving birth.

Working animals

- behavioural traits, for example, retrieving/chasing/rounding/obedience in relevant dog breeds
- sense of smell (tracking/rescue dogs)
- fitness, strength, movement, speed (sport horses).

Avians

In birds, it is also true that their role is what determines their desirable characteristics:

Meat production

- muscle mass
- fertility.

Egg production

- regular laying ability
- long period of laying.

Pets

- particular chosen colourations
- feather/beak/feet/song quality
- disease resistance
- fertility.

Survival strategies

Survival strategies that help to ensure success of different taxa include: numbers of offspring, extent of parental care, development at birth, growth rate, hardiness and defence mechanisms. In your assessment you would need to consider and research the specific strategies for the species given.

> Parent birds will be selected for frequently producing a good clutch of eggs.

Herptiles

Breeders may try to obtain offspring with bright colours and different **morphs** in order to appeal to buyers. Breeders may also breed from animals that are well socialised, in good health and that have adapted well to captivity, because their offspring are more likely to do better in a captive environment than those bred from wild-caught **herptiles** (reptiles and amphibians).

European Studbook Foundation (ESF)

The **ESF** is a non-profit organisation that promotes and manages **studbooks** of reptiles and amphibians in captivity. This helps to maintain a genetically healthy breeding programme of reptiles and amphibians.

Pedigrees

A **pedigree** is a record of the phenotypic background of an animal, its siblings, parents, grandparents and previous generations. Often 'purebred' animals (which result from generations of offspring from a recognised breed) will have a pedigree certificate. Some aspects of inheritance can be tracked in pedigree animals. Pedigree records or charts include dominant and recessive patterns of inheritance, which could enable the probability of inheritance to be calculated (for example, of a disease).

> Consider the advantages of animals that are sociable and docile when in a breeding scenario.

Now try this

1 Explain why breeders select animals that are sociable/docile.
2 Explain why breeders use animals with unusual or special markings/colouration.

Construction of breeding programmes

Key factors are evaluated in **breeding programmes** to identify the best animals to breed to produce the desired outcome.

Estimated breeding values

The breeding value of an animal is how **desirable** its genes are, as estimated from its phenotype. Each animal is given a score of breeding value for a particular trait, normally in the same units as the measurement of the trait where possible. For example, a bull with an **EBV** (estimated breeding value) of +50 kg for 600-day weight is estimated to have genetic merit 50 kg above the breed base of 0 kg. Estimated breeding values are published by relevant organisations for a range of factors and species.

Beef cattle traits

Traits evaluated for estimating breeding values in beef cattle can include:

☑ **200-day growth:** high positive values are preferable as they indicate a good potential for initial growth.

☑ **Birthing ease:** high values are preferable as there is less chance of complications during parturition.

☑ **Gestation length:** low values can be preferable as birthing is likely to be easier, but too low and the offspring may not survive.

☑ **Carcass weight:** indicates the potential meat yield.

☑ **Rib and rump fat:** indicates the potential for lean meat production.

Inbreeding coefficients

The **inbreeding coefficient** (F) gives the probability of how genetically similar (closely related) two individuals are. A higher coefficient (shown as a percentage) means the individuals are more closely related.

For example:

Parent–offspring mating: 25%
Full sibling mating: 25%
Half sibling or grandparent/grandchild mating: 12.5%
First cousin: 6.25%

(These would be higher if there is previous inbreeding in earlier generations.)

Inbreeding coefficient examples

The two remaining wolves on Isle Royale National Park, USA, are so closely related that any potential offspring of the pair are estimated to have an inbreeding coefficient of 43.8%.

According to Dr John Armstrong, author of *Inbreeding and diversity* (2000), standard poodles with an inbreeding coefficient of less than 6.25% have 4 years longer life expectancy than standard poodles with an inbreeding coefficient over 6.25%.

Selection definitions

1 **Selection differential:** the difference between the average population and the average of the animals selected to breed. For example, a flock of sheep has an average weight of 31 kg and the animals selected to breed have weights of 33 kg and 39 kg. The average mass of animals selected to breed is 36 kg, which gives a selection differential of 5 kg.

2 **Selection intensity:** the superiority of the breeding animal's characteristics compared to the rest of the population. It looks at how consistent the selection differential is within the animals selected to breed.

3 **Selection response:** the genetic gain between generations (how much the population is considered to be improved).

Now try this

The Department for Environment, Food and Rural Affairs (DEFRA) needs to produce some resources to promote the benefits of effective mating schemes.

Explain **two** ways that improvements to animal welfare and benefits to humans can be gained through the use of animal selection.

Factors affecting selection for breeding

Breeders need to consider various factors, relevant to the species, when deciding which animals are suitable for breeding programmes.

Large heads may prove a problem during parturition.

For example, levels of hormones or colourful plumage in birds (desirable in the pet trade).

Birds and herptiles with bright colours or unusual markings are more attractive to customers.

Factors affecting selection for breeding

- Head shape/size
- Size/condition of teats
- Sexual characteristics
- Markings/colour of skin/coat/feathers
- Condition of mouth/beak/teeth/eyes
- Size/condition of limbs
- Body score
- Condition/conformation of limbs/claws/hooves/feet
- Desirable characteristics
- Health
- Posture/conformation
- Horns
- Temperament
- Anus/cloaca/genitals
- Previous breeding history
  - ease of parturition
  - egg laying frequency
  - incubation requirements/duration
  - fertility
  - mortality rates
  - maternal/paternal behaviours
  - lactation
  - multiple offspring

Make sure you consider any sexual dimorphisms present in any species you are considering, including in your set task.

A heifer is a cow that has not yet given birth to a calf.

Assessing suitability

Both males and females should show desirable characteristics, as well as a good breeding history if they have been bred previously. For example, a **heifer** or cow that has good physical characteristics (weight/udder condition/hooves/limbs) with a successful breeding history would be a good candidate. This also includes the behaviour/temperament of the animal.

🔗 **Links** See page 13 for more on behaviour assessment.

Lameness and feet problems in cows can be hereditary. Cows with these problems would not be selected for breeding.

Now try this

What would a breeder or farmer be looking for when assessing the suitability of a heifer to be put into a breeding programme specifically for dairy cattle?

Use of information in animal evaluation

Different breeds of a species, or individuals within a breed, may need to be evaluated for various reasons.

Breed profiles

The general information found in **breed profiles** helps breeders and potential owners assess whether that breed is suitable for their needs. For example, in choosing a pet dog, a dog breed requiring lots of exercise may not be suitable for an owner who cannot take them out regularly. They can also indicate whether a breed is suitable to be trained as a working dog, or whether it is a breed associated with health problems that could be costly to treat.

Breed profile contents

These include:

- life span
- breed-specific health conditions
- temperament
- general care
- special care
- potential adult size (height, weight)
- dietary needs.

Uses of breed profiles

- In the **pet industry**, breed profiles can help potential owners make informed choices. This will reduce the likelihood of animals ending up in rescue centres or suffering.
- Food producers can select breeds developed for specific **food production** purposes so that profit margins and welfare standards are maximised.

Evaluating individual animals

Within a specific breed, animals may need to be evaluated to:

- establish which animals are fit to be sent to market or slaughter
- select the best animals to breed
- choose the best animal for showing
- ascertain those most likely to be successfully trained.

🔗 **Links** For more on individual animal assessment, see page 12. For more on factors affecting selection for breeding, see page 10.

Example of a breed profile

Breed	Dalmatian
Information	Dalmatians are well known for their spotted coats. Historically, Dalmatians were used as guard dogs. They were also associated with fire fighters, running ahead of the fire engine when it was pulled by horses, clearing the way and making sure the horses stay calm.
Life span	10–13 years
Litter size	6–9 puppies
Size	Medium size, ranging from 48–60 cm tall, broad and strong musculature
Weight	16–32 kg, with males being heavier
Colour and coat	Black and white, liver and white; short, fine hair
Temperament	Active, intelligent and sociable
Known potential health issues	Urinary problems impacting on the kidneys and liver, hereditary deafness, skin problems

A dalmatian.

Now try this

Images and diagrams may help display visually some characteristic of the animals within the breed profile.

1 Why are breed profiles useful?
2 What important aspects should be detailed within a breed profile and why?

Animal condition

Different techniques are used to assess the age and condition of individual animals, including **body condition scoring** and **tooth condition**.

Body condition score (BCS)

BCS is a common technique used by breeders and pet owners to measure whether or not their animals are in **good health**. Healthy animals with no welfare issues have a higher chance of being fertile, and having fewer complications if used to breed from.

BCS:

- monitors condition – could indicate disease/ill health or nutritional deficiencies
- minimises breeding difficulties
- can indicate other problems, such as incorrect management of herd/stock.

Body condition scoring in pigs

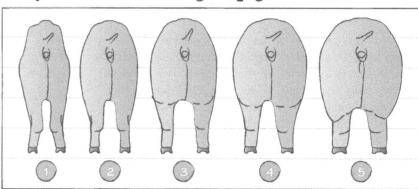

Score	Condition	Detection of ribs, backbone and pelvis
1	Emaciated	Obvious
2	Thin	Easily detected with pressure
3	Ideal	Barely felt with firm pressure
4	Fat	None
5	Overly fat	None

Body condition scoring is used by breeders in a range of species, for example, pigs.

Tooth condition

An experienced handler or vet looks for worn teeth, the number of teeth present and the build-up of tartar.

Factors that can affect tooth condition:

1 **Age**: in mammals, the condition of teeth can indicate the age of an animal.

2 **Diet**: soft, fatty diets can cause decay or tartar build-up, for example, wet cat food. The teeth of grazing animals may be affected if they are not allowed to graze for sufficient time.

3 **Illness**: disease or health conditions may cause mouth or gum issues, causing decay or tooth loss.

4 **Enrichment**: in some species, teeth grow throughout their lifetime, for example, rats and rabbits. They should be offered objects to allow their teeth to be worn down to avoid overgrown or abnormal teeth.

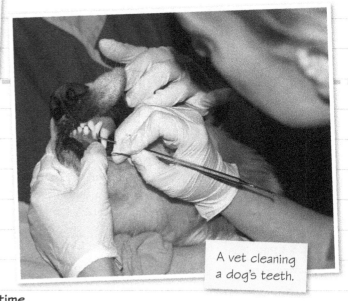

A vet cleaning a dog's teeth.

You can describe the animals' husbandry routine that may have an impact, for example, their diet. Are they grazing animals that have been allowed to graze sufficiently, or dogs that are provided with wet food?

Now try this

Describe **three** factors that can impact on the condition of an animal's teeth.

Behaviour assessment

Behaviour and temperament assessments are important in deciding whether or not animals should be included in a breeding programme.

Behaviour assessments

Assessing the **behaviour** of animals helps to maintain the safety and welfare of all animals, owners and handlers involved in a breeding programme. Appropriate behaviour also contributes to successful mating and rearing of young.

Important behaviours to assess include **social behaviour** of both males and females, **reproductive behaviours** once male and female are brought together, and **parenting behaviours**.

For example, cattle breeders will choose well-socialised cattle, to minimise aggression, and those who show a desire to mate. Strong maternal behaviours are desirable after parturition.

Temperament

Temperament of the animals involved is also important. Breeders would consider:

- **sociability with other animals** – pets and animals used in food production are often kept in social groupings
- **sociability with people** – to ensure ease of handling
- **fear or aggression** – indicates the likelihood of fights or injuries occurring between animals
- **reactions to different environments** – animals will be exposed to different types of housing and other environments when being bred.

Behavioural considerations in conservation programmes

In **conservation programmes**, animals are often not socialised with humans or other animals. Measures can be taken to assist safe and successful breeding, such as:

- using equipment such as crush cages for artificial insemination
- slowly introducing animals to mates in natural reproductive programmes
- monitoring the animals' behaviour and intervening if necessary (for example, fighting or poor maternal behaviour).

Animals showing poor maternal or paternal behaviour are less likely to be selected for future breeding programmes.

It is important that animals to be returned to the wild do not have too much contact with humans as this can have an impact on their reintroduction and/or rehabilitation.

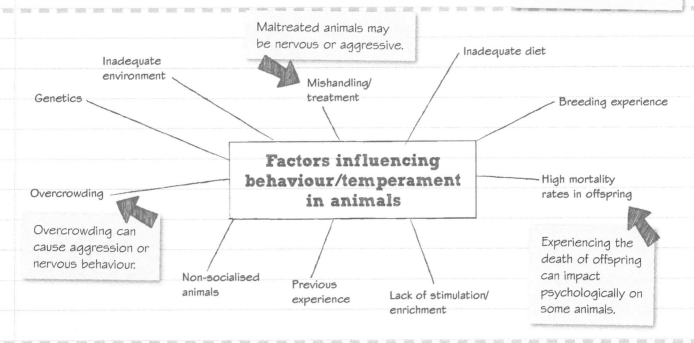

Maltreated animals may be nervous or aggressive.

Inadequate environment

Genetics

Inadequate diet

Breeding experience

Factors influencing behaviour/temperament in animals

Overcrowding

Overcrowding can cause aggression or nervous behaviour.

Mishandling/ treatment

High mortality rates in offspring

Experiencing the death of offspring can impact psychologically on some animals.

Non-socialised animals

Previous experience

Lack of stimulation/ enrichment

Now try this

1 Discuss why it is important to conduct a behaviour assessment on animals before breeding.
2 List **three** undesirable behaviours for potential breeding animals.

Legislative requirements for breeding animals

The **Animal Welfare Act 2006** is the main legislation to follow when dealing with animals. Other legislation applies to breeding in scientific research and for dogs. You will also need to know about this for Unit 3, see page 115.

The Animal Welfare Act 2006

This act places a **duty of care** on animal owners, breeders and farmers to ensure that the animal's basic needs are met.

Guinea pig eating.

Five welfare needs

The **five welfare needs** are shown below. These vary with life stage. For example, during pregnancy and lactation a female animal will have differing nutritional needs, behavioural patterns and need more careful handling and a stress-free environment.

Owners should consider size and space of accommodation, including height.

Welfare needs

1 Need for a **suitable environment**

2 Need for a **suitable diet**

Owners should ensure sufficient types of food and water are available.

3 Need to be able to **exhibit normal behaviour patterns**

Opportunity for normal behaviours should be allowed, for example, providing nesting material.

5 Need to be **protected from pain, suffering and disease**

For example, enclosure should be safe for the animal and animal should be vaccinated.

Social and solitary species should be provided with that opportunity.

4 Need to be **housed** with, or apart from, other animals

Animals (Scientific Procedures) Act 1986

The ASPA applies to all vertebrates and cephalopods. Animals must be obtained from a designated breeding establishment. An **establishment licence** must be in place before animals can be bred. Suitably qualified personnel are responsible for the care and welfare of the animals (for example, an animal technician needs to be with an overseeing veterinary surgeon). Animals can then be bred on demand to be used in projects that again have all the licences in place.

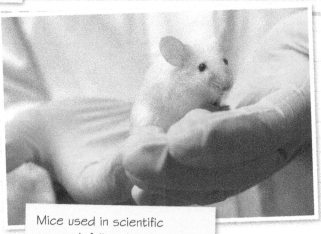
Mice used in scientific research fall under the ASPA.

Now try this

Explain how the five animal needs are linked to the breeding of animals.

Try and think of one answer for each need.

Legislative requirements for breeding dogs

There is specific legislation that applies to the **breeding** and **selling** of **dogs**.

Sale of dogs

Dogs can only be sold by a **licensed breeding establishment**, or a **licensed pet store**. The puppies sold must be at least **eight weeks** old, unless being sold to another licensed pet store or breeding establishment. Any dogs sold must be wearing an **identifying badge** or tag, displaying the licensed breeding establishment it was born at.

The law

The **Breeding and Sale of Dogs (Welfare) Act 1999**, along with the **Breeding of Dogs Acts 1973** and **1991**, are pieces of legislation that apply to people who wish to keep breeding stock and sell the pups. A dog licence from the local council is required by anyone who:

- breeds and sells dogs, irrespective of how many litters are produced
- produces over five litters of pups in 12 months, even if they do not sell the pups.

Legislation requirements

The legislation focuses on how the dogs are kept, sold and transported. The breeder must ensure the day-to-day care of the dogs, including suitable accommodation and feeding and exercise, and they must keep records that show they are meeting all required conditions.

A few examples of the conditions that must be met include:

1. Dogs must be kept in suitable accommodation, for example, size, number of dogs in the kennel, exercising facilities, temperature, lighting, ventilation and cleanliness.

2. Dogs must be provided with correct food, drink and bedding material, and must be adequately exercised, and visited at suitable intervals.

3. Disease control must be in place, for example, high levels of hygiene and dogs vaccinated. Bitches must:
   - be over one year old before breeding
   - produce fewer than six litters in their lifetime
   - have only one litter every 12 months.

Bitches must be over one year old to have puppies.

Now try this

Explain the key factors related to the breeding life of a bitch when following the Breeding and Sale of Dogs (Welfare) Act 1999.

Handling techniques and strategies

Handling techniques need to be appropriate for different animals according to their species, size, temperament, health status or life stage.

Considerations

Different handling techniques or strategies will be appropriate for:

- young animals, including recently hatched
- animals that are due to give birth
- animals that have offspring with them, for example, lactating mothers
- senior animals (experienced animals may be used as care givers or for socialisation)
- animals with specific health conditions.

Each technique will have its own good and bad points in relation to the welfare of the parent, offspring/foetus/eggs and the environment.

Equipment

Species-specific equipment is used to make handling easier and less stressful during a health check of adults or offspring, artificial insemination or the introduction of animals that are due to be bred together. **Crush cages** may be used on larger breeding stock, such as cattle, when assessing whether animals are fit for breeding, or to support the artificial insemination process. **Breeding cages/boxes** may be used to ensure a regulated temperature and environment for the breeding animals.

Examples of suitable handling techniques and strategies

Is it the correct equipment for the species/scenario/task in hand?

Will it help reduce stress?

Is it safe?

Equipment must be assessed before being used

Will it reduce the need for physical contact?

Is it in good condition?

Small mammal handling techniques	Large mammal handling techniques	Herptile and avian handling techniques
Scoop method: protect and support full body weight, especially if animal is pregnant. **Scent hands:** to ensure young animals do not smell of other animals/humans.	**Approach:** quiet and slow reduces stress in all animals. **Equipment:** animals may need to be moved to a smaller area for inspection, for example, using hurdles to enclose them.	**Scoop method:** supports full body weight and offers protection, especially if **gravid**. **Avoid handling:** these animals can be very sensitive and easily damaged (especially eggs).

Farrowing crates allow for piglets to nurse whilst minimising the mortality rates of the offspring. However, the RSPCA are concerned about farrowing crates as the sow is confined and cannot build a nest. They have been banned in Sweden, Norway and Switzerland.

Gravid means carrying eggs or young.

Now try this

State **two** advantages and **two** disadvantages when using the scoop handling method.

Consider which species and ages you may use this technique on.

Preparation for breeding

Incorporating an animal's natural reproductive strategies into a breeding plan can increase breeding success.

Preparation for mating

Planning ahead will help reduce the likelihood of disease and complications with conception. If the animals are in the correct conditions, there will be a higher chance of success.

It is important to manage the females selected for breeding to ensure their welfare is met. Key things to consider are:

- How often should the female be bred from?
- How long should be left between breeding for the female to recover?
- What is the minimum age the female should be bred from?
- What is the maximum age the female should be bred from?
- How many times should the female be used to breed?

Considerations for breeding

Considerations may include:

✓ **age of the animal**

✓ **oestrous/natural breeding season:** introducing animals outside of mating season, or when not receptive, can cause conflict and injury

✓ **animal behaviour:** do the potential breeding pair interact well together?

✓ **suitability of location:** for mating or artificial insemination

✓ **health:** are the animals in general good health and free from diseases that could be sexually transmitted?

✓ **breeding history:** previous success or failure.

Mate recognition systems

Mate recognition systems are a set of characteristic signals that indicate the suitability of another animal as a potential mate. These **natural reproductive strategies** can be exploited by breeders to encourage a successful mating. Breeding plans should allow sufficient time for animals to exhibit these behaviours and choose a preferred mate. Animals that do not mate, or do not produce viable offspring within a set time frame, should be removed from the breeding programme.

Auditory – calls can signal species identity and readiness to mate, e.g. different species of bird have different calls which help identify them as a particular species and their suitability as a mate.

Behavioural – mating behaviours allow animals to demonstrate their suitability as a mate, e.g. deer will fight for the position of dominant stag, which then gets to mate.

Mate recognition systems

Visual – in some species there are clear differences between males and females. Some animals use their appearance to attract a mate, e.g. male peacocks display their tail feathers to attract a mate.

Olfactory – pheromones can be species specific and link to an animal's fertility, e.g. female snakes release pheromones to indicate their fertility.

Now try this

A zoo keeper wants to create a breeding plan for Humboldt penguins.

Research the mate recognition strategies these penguins use, and how they can be incorporated into the breeding plan.

Reproductive technologies

Reproductive technologies can be used to monitor and regulate breeding. They can be used individually or in combination. By regulating breeding, closer monitoring of the animals can take place and more accurate planning for birth can be done.

Identifying ovulation in animals

Being aware of **ovulation** and enabling mating to occur at that time increases the chances of breeding success. It can be identified by:

1 detecting body temperature increases. This indicates the onset of ovulation and the time for successful breeding.

2 detecting decreases in the pH of the vulva using a pH test strip or electronic probe.

3 identifying hormone changes. A vet will take a blood sample from the animal and have the levels of reproductive hormones analysed.

4 observing behavioural changes, such as:

- allowing other animals to mount her – marking harnesses or pressure-sensitive pads will identify which animals have been mounted
- increased socialisation with other animals
- vocalisation
- restlessness – pedometers attached to an animal's foreleg can indicate increased activity levels
- noticing physical changes, for example, swollen vulva with excess discharge.

Hormone therapy

Hormones can be used to affect ovulation in two ways:

1 **Superovulation:** to increase the number of ova (egg cells) the female produces with each oestrous cycle. If the female produces more eggs, there is more potential for mating to lead to offspring and for higher numbers of offspring to be born.

2 **Synchronisation:** to regulate the oestrous cycle of a group of animals. Offspring will then be due at a similar time, making monitoring of the animals quicker and easier. It can also be used alongside **artificial insemination**.

In animal collections and breeding programmes, artificial insemination and in-vitro fertilisation are commonly used. This allows males used in reproduction to be selected from a wider area, reducing safety issues by not bringing animals into close contact with each other (which can also increase the chances of a successful mating) and giving more control over the time of conception.

Artificial insemination: Sperm collected from a male with desirable traits is used to manually inseminate the female. This enables a male with desirable traits to produce many offspring. Semen collection and sale is a lucrative commercial business.

Reproductive techniques

In-vitro fertilisation (IVF): An embryo is created in the lab from sperm and eggs from parents with desirable traits. The embryo may be implanted into the mother's womb, or in an unrelated female.

Sperm sexing: Spermatozoa carrying male or female genes can be separated and used with IVF to influence the proportions of male and female offspring produced. This is commercially important for both milk production (preference for female offspring) and meat production (preference for male offspring).

Embryo transfer: Superovulation is used to increase the number of ova in the donor, who is then inseminated repeatedly over a few days. The donor's uterus is flushed before the fertilised eggs can implant and the eggs collected. The eggs are analysed and the best transferred to the recipient's uterus. This is used to increase the number of offspring produced from a female with desirable characteristics.

Links Reproductive technologies that involve genetic engineering can be found on page 33.

Now try this

A dairy farmer wants to increase the proportion of female calves born in his herd.

Identify the reproductive technologies that could help the farmer do this.

Contraception

There are various reasons why we may want to control an animal's fertility to prevent reproduction, and there are many methods of contraception.

Purpose of contraception

Contraception is important for the following reasons:

- population control
- control of the timing and frequency of pregnancy
- managing or changing sexual behaviours (for example, spraying, crying, roaming)
- control of aggression
- avoiding the mess associated with being on heat in domestic pets
- preventing some health problems
- pest control.

Selecting a contraceptive method

Contraception needs to be carefully chosen, taking into account species differences in physiology and life history, and its purpose. For example, progestin (hormone)-based contraceptives are widely used in **ungulates** and primates but may cause uterine diseases in dogs.

Ease and cost need to be considered. For example, a contraceptive feed would be most appropriate for wild or feral animals, rather than surgical methods.

Either a permanent or reversible method may be required.

Surgical contraception

Surgical contraceptives involve the removal of reproductive organs, or removing the blood supply to the organs to prevent them functioning. They are permanent.

Female
Neutering: ovaries and/or uterus are removed via abdominal surgery under anaesthetic.

Male
Castration: testes are surgically removed under anaesthetic.

Banding: (e.g. in sheep) a band is placed at the base of the scrotum to restrict blood supply, causing the testes to drop off.

Burdizzo pincher castration: (e.g. in cattle) a clamp is used to cut the blood supply leading to the testicles.

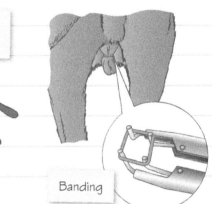

A Burdizzo pincher

Banding

Non-surgical contraception

Hormonal, chemical or immune-contraceptives may be administered by injection, in feed or in implants inserted under the skin.

	Description	Permanent or non-permanent
Hormonal contraception	Uses **synthetic reproductive hormones** to suppress ovulation or **hormone antagonist** drugs that block the action of reproductive hormones.	Non-permanent. Requires carefully-timed repeated oral doses, but long-acting injections and implants are available. It can also be added to feed.
Contraceptive vaccines	Vaccines against egg or sperm proteins or against hormone receptors. These new methods are known as **immune-contraception**.	Non-permanent. Requires repeated injections.
Chemical castration	Chemicals injected into the testes to make them inactive.	Permanent

Now try this

Consider the welfare implications of using the banding castration method on livestock.

Pregnancy diagnosis in mammals

Pregnancy diagnosis is the confirmation that an animal is pregnant. There are several methods, including visual signs, as well as clinical diagnosis and laboratory tests.

Why is pregnancy diagnosis important?

1 **Husbandry management** plans can be implemented to ensure the correct care is provided.

2 Early diagnosis can also be **cost effective** – knowing which males have bred successfully means they can be used again quickly within breeding programmes. This is especially important in zoo breeding programmes.

Hormone testing

Changes in **progesterone** and **relaxin** occur after **conception**. These hormones can be detected in blood or urine and used to diagnose pregnancy. Home testing kits involving small blood samples or urine are available, but seeking veterinary attention and confirming the test in the laboratory is advisable. In cows, progesterone increases around 24 days after conception, and in dogs progesterone is at its highest around 20 days after.

Ultrasound and palpation

Clinical examination or palpation, with **ultrasound**, will reveal whether an animal is pregnant. An ultrasound in dogs, for example, will detect pregnancy after Day 25 of gestation.

Palpation is physically examining the animal to establish the presence of a foetus. Rectal palpation is used for large animals and external palpation of the abdomen for small animals.

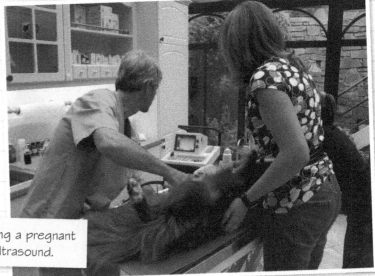

A vet checking a pregnant bitch using ultrasound.

Inactivity or restlessness

Increase in udder size or teats

Increase in weight

Visual signs of pregnancy

Change in appetite

Distended abdomen

Vaginal discharge

Animal does not return to oestrous

Visual signs of pregnancy

Visual signs can often indicate whether animals are pregnant. In dogs, for example, in a bitch that has been mated and is progressing with her pregnancy, you would start to see teats enlarge and become more 'pink'.

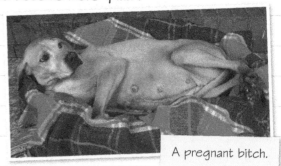

A pregnant bitch.

Now try this

Give **three** signs that would indicate a dog is pregnant.

 Look at the spider diagram and photo above. You can describe the obvious changes that can be seen and more discreet changes such as behaviour.

Preparing for parturition

It is important to monitor for the signs of parturition and prepare for parturition.

Signs of parturition

- Behaviour changes – **restlessness**, **straining** or **vocalising**
- Animals may **isolate** themselves from the rest of the group
- Feeding patterns may change – eating less around the time of parturition
- Vulva may become **swollen** and soften
- Cervix will **dilate**
- **Mucus** will be seen around the vulva
- Body temperature may change
- Amniotic sac will break causing **amniotic fluid** to be produced from the vulva
- **Milk production** may be evident.

All animals are different and may show signs of parturition in any order and to different extents. In some animals, parturition will progress rapidly after the first signs and others can take a lot longer.

Signs of parturition in sheep

☑ **Swelling of the vulva:** can occur 4–5 days before birth.

☑ **'Let down' of milk:** 48 hours prior to birth is usual.

☑ **Panting/increased respiration:** usually begins 24 hours before birth.

☑ **Restlessness:** usually begins 24 hours before birth and continues until about an hour before birth.

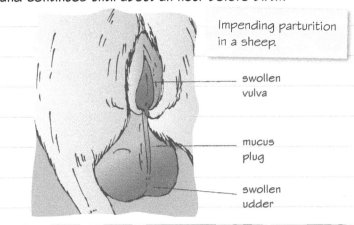

Impending parturition in a sheep.

swollen vulva

mucus plug

swollen udder

Preparations

Equipment will need to be gathered, and location of the animals may need changing.

Preparations for managing the stages of parturition will depend on the species and any breeding history. If the female has previously experienced any birthing problems additional measures may need to be implemented. These could be extra **veterinary check-ups** and monitoring, for example, checking of calcium levels.

Equipment

Towels may be needed to keep the offspring warm, as well as heat lamps or mats. **Iodine** can be used as an **antiseptic** for livestock to improve the hygiene of anyone assisting and to promote health in the offspring. In lambs, the remains of the **umbilical cord** should be covered in iodine to reduce the chance of infection through the cord.

There is always the risk of unplanned emergencies during birth, so it is a good idea to have a set of portable essential equipment to hand.

Environment/location

Animals may need moving to a quiet, **isolated** area, with suitable access for handlers, for example, moving mares into a foaling stable.

Fresh bedding should be provided for some animals, so the mother can create a **nesting area** for the offspring. **Water facilities** should be suitable for the birthing area, for example, ensuring that offspring cannot climb in the water tray and drown, or get pneumonia.

Now try this

You are completing a work placement on a farm with a small dairy herd and calving season is approaching. You are working with the herd manager to ensure the farm has completed all the necessary preparations.

List the types of preparations that need to be made, including practical knowledge which would be useful.

Parturition

Ideally, parturition occurs naturally, but handlers should be prepared to intervene safely if required.

Parturition in live-bearing animals

In live-bearing animals, there are distinct stages of parturition:

1 **Preparatory stage:** dilation of the birth canal.

2 **First expulsion stage:** expulsion of the foetus.

3 **Second expulsion stage:** expulsion of the placenta.

1 Cervix widens (dilates) and uterus begins contractions.

2 Foetus passes through the cervix, pelvic region and then the vagina.

3 Uterine contractions expel membranes.

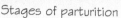
Stages of parturition

Problems

Problems during parturition could include:

- offspring getting stuck in the birth canal
- **breech presentation**
- **haemorrhage**.
- **uterine prolapse**
- **eclampsia**

Veterinary attention should be sought if any of these are observed.

Monitoring and handler involvement

Handlers should remain unobtrusive and should not intervene with the birth, unless serious health issues are observed. Examples are **malpresentation** (abnormal position of the foetus) or **lack of response** from the offspring. Interference can increase the mother's stress levels, making the birth more complicated, and can have a negative effect on the initial **bonding** between the mother and her offspring. **Monitoring** should continue until the expulsion of the placenta is complete.

Health and safety of breeders or technicians

A **risk-assessment** should be carried out of all aspects of working with animals. This document should be available to personnel, highlighting any **hazards** that may occur, how likely they are to happen and offering appropriate solutions that mitigate the risk.

Common health and safety problems include:

☑ spread of **zoonotic** diseases

☑ attack by an animal

☑ allergies.

Zoonotic diseases (zoonoses) are diseases that can be spread between animals and humans.

Personal protective equipment (PPE)

Using PPE reduces the risk of **zoonoses** and the spread of diseases between the animals, for example, tuberculosis can rapidly spread between cattle. PPE includes:

- disposable aprons
- disposable gloves
- face masks
- waterproof footwear
- steel toe-capped footwear.

Any equipment or PPE used should be **unscented** so that is does not affect animal-to-animal bonding in the mother and offspring.

Now try this

You have noticed your heavily pregnant ewe is preparing to give birth.

Detail what equipment may be necessary during this time.

Housing of breeding animals

Various factors need to be considered and accommodated in the housing of breeding animals.

Housing design requirements

Any housing for breeding animals needs to fulfil the following requirements:

- space to incorporate mother and offspring comfortably
- safe access for handlers
- accommodate changes in the behaviour that may be seen during pregnancy and lactation
- appropriate temperature
- sufficient ventilation
- enrichment that allows the mother to prepare for birth.

Examples of room/water temperatures

Different species have different temperature requirements when breeding:

Pigs	24–30°C
Guppies	23–28°C
Budgies	15–20°C

Considerations

In order to plan housing for breeding animals, you need to consider the following:

Species – size and litter.

Breeding status – during parturition, during pregnancy or to be mated.

Lactating – additional facilities required to help with feeding and weaning.

Behaviour – animals may behave differently during pregnancy. For example, they may start nesting or become more aggressive.

Laboratory-style racks are used to house rodents that are both breeding and non-breeding. Extra enrichment, for example, nesting material, is supplied to the breeders in order to help them prepare for parturition.

Access – a wire lid is preferable to view the rodents without being too invasive.

Changes in the behaviour – rats may sleep more before parturition, as well as become more territorial or aggressive. Housing should consider this and not lead to unnecessary stress for the animals.

Space – 2 cubic feet per rat, but more is better.

Housing for breeding rodents (e.g. rats)

Enrichment – proved nesting material such as shredded paper or tissue, and shelters, could be provided, as well as gnawing material.

Temperature – ideal temperature is 19–23°C.

Ventilation – the wire lid is sufficient ventilation, but there should be no draughts.

Now try this

Why is knowing about normal animal behaviour during pregnancy and parturition important?

Care plans

Careful plans need to be developed to cover all eventualities during the birth of animals, and care of the mother and offspring up until weaning/independence. The following lists give points to consider.

Care during parturition or hatching

Plans should include the following questions:

- What environment/housing is required?
- What is the typical duration of parturition or incubation?
- What equipment should I have ready?
- What do I need to do to reduce the chance of infection?
- What care may the mother require, and am I prepared?

Care considerations, plans and monitoring, and record keeping for a flock of sheep on a hillside will differ greatly to those of a dog in a family home or chickens in a meat production facility.

Housing of newborn/newly hatched animals

You should consider the following:

- Is any specialist equipment required to separate mother and animals?
- Is extra heating required?
- Is water in a suitable and safe container for young?
- Has early suckling taken place, as it is vital that mammals receive colostrum within the first 12 hours?

Baby goats

Being prepared for complications

You must:

- ensure you are aware of signs of distress/ complications to both mother and offspring
- have the emergency vet's number to hand
- understand how to care for a mother who has required a caesarean section
- know how to deal with any still born offspring, including disposal
- have colostrum powders and feeding equipment ready in case of death of the mother or maternal rejection of the offspring
- know how to encourage a reluctant mother to suckle
- know how to encourage adoption of orphaned animals by a new mother.

Precocial means being active and able to move freely from birth or hatching.

Nutrition for precocial and orphaned animals

Ask yourself the following questions:

- What type of feed is required at which stage of development?
- Are all offspring receiving adequate feed?
- Are other elements required for successful nutrition, such as grit or ensuring the gut is colonised with bacteria?
- Is there a plan for feeding orphaned offspring, such as a suitable diet and is the correct equipment available?

Weaning

- Will weaning occur naturally or will it be enforced?
- When should it happen?

For some species there are legal requirements for record keeping.

Monitoring and record keeping

- What information and observations about mother and offspring should be recorded at the different stages?
- How are individual animals identified?
- What method of recording is most suitable for the situation?

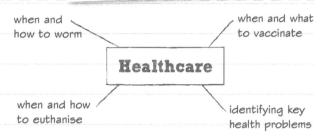

when and how to worm — when and what to vaccinate

Healthcare

when and how to euthanise — identifying key health problems

Now try this

Explain why dietary requirements for both the offspring and mother should be included in care plans.

Care plans: dogs – parturition and neonatal care

All eventualities should be planned for when preparing to care for a whelping bitch and her pups.

Preparations

A high-sided disposable or easily sterilisable **whelping box** should be supplied to the dam to keep her and her pups together, safe and warm. Also ensure you have the following available:

- towels
- contact details for the vet in case of emergency
- heat pads or lamps
- small blunt scissors
- equipment and formula milk for hand rearing.

Normal whelping

☑ Panting and straining is usual.

☑ Parturition lasts 3–12 hours, depending on the number of pups to be delivered.

☑ 20 minutes–2 hours between the delivery of each pup.

☑ A placenta should soon follow the delivery of each pup, which the dam may ingest.

☑ Once all pups have been delivered, the dam will relax and commence care-giving behaviours.

Dams should be monitored for complications. Intervention may be required if the dam does not continue to deliver after two hours.

Complications

Complication	Next step
Restless dam	Ensure the dam has access to a quiet comfy area. Contact vet if this continues.
Umbilical cord	Cut the cord with blunt scissors if the mother does not chew it.
Pup stuck in birth canal (visible)	Use a clean towel/tissue and gently hold onto the pup, a gentle movement may start contractions again. If it remains stuck, contact the vet immediately.
Excessive bleeding	Contact the vet immediately – the dam could be suffering from a ruptured uterus.
Stillbirth	Remove whilst the dam is distracted and have cremated at the vets.
Caesarean section	Feed only small amounts at first to ensure no vomiting occurs; limit movement; help pups to nurse; increase observations.

Care of newborn puppies

The pups will remain in the whelping box with their mother. Puppies may need help with:

- **suckling** – monitor each puppy's intake, they may need moving closer to the teat.
- **colostrum** – this is required in the first 24 hours, so hand rearing may be required if the pups cannot latch on.
- **hygiene** – clean the puppies' umbilical cords in the first 24 hours using alcohol-based animal-friendly products.

Records

A breeder will want to record information relating to birth to ensure the health of all dogs, as well as decide whether the dogs should be bred again.

Duration of parturition	The process should be recorded, as well as the time between each pup
Complications	Complications should be recorded
Maternal behaviour	Good/poor maternal behaviours should be documented, e.g. licking, chewing umbilical cord
Pups	Should be identified and weight regularly monitored

Make sure you understand maternal and paternal care of neonates and the strategies used by both mammals and birds. The information on pages 25–28 is for dogs and chickens as examples.

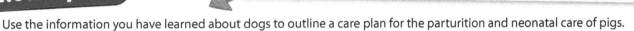

Now try this

Use the information you have learned about dogs to outline a care plan for the parturition and neonatal care of pigs.

Care plans:
dogs – early life to weaning

Preparations should be made for the **early care** of pups before whelping.

Housing

The puppies remain in the **whelping box** until around three weeks of age, when they can move around more freely. A large contained area should then be given to allow the puppies more **exercise space**. They can also start to explore outside in areas that are not accessed by other dogs, especially unvaccinated ones.

Healthcare

Vaccinations should be given against canine parvovirus, canine distemper virus, leptospirosis and infectious canine hepatitis at 8 and 10 weeks. Puppies should not be walked until 1–2 weeks after the second vaccination. **Worming** should be carried out every 2–3 weeks from two weeks of age until they are around 12 weeks.

Records and monitoring

Weight should be recorded at least weekly in the first few months and development monitored.

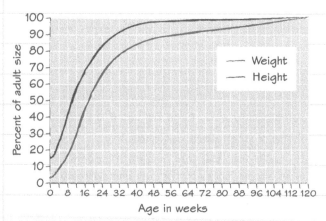

A graph showing the typical growth pattern for puppies.

Hand rearing

Equipment	Syringes, bottles, teats
Frequency	Every four hours
Technique	Feed pups lying on their stomachs
Duration	Three weeks
Weaning	Offer milk and food in a bowl

Puppies will start to eat moistened solid foods from around 3–4 weeks, whilst continuing to feed from their mother. They should be **weaned** around **eight weeks**, and preferably left together until around 10 weeks.

Puppy feeding from bowl.

	Stages in early development of puppies
Eyes	Open around two weeks (8–13 days)
Ears	Open around two weeks (13–20 days)
Dentition	Start to appear around three weeks; 'baby teeth' all through by Week 8
Activity	Become more active around Week 2

Common problems

👎 **Gastric infection** can lead to **malnutrition** and dehydration, so should be carefully managed.

👎 Puppies are susceptible to infectious diseases such as **parvovirus** until they have been vaccinated.

Puppy being vaccinated.

Now try this

Prepare a feeding schedule for orphaned puppies.

 Consider what the puppies require in each week of its early stages.

Care plans: poultry – egg incubation

Care of poultry within a commercial meat production unit will be very different to that of a backyard breeder.

Natural incubation/ nesting

Some eggs can be placed under a sitting hen for **incubation**. The hen will rotate the eggs, leaving them only for short periods when she eats and drinks. Plenty of food and fresh water should be available for the hen.

During this time, the hen will fight off any **threat** she perceives towards the eggs (including people).

Natural incubation can take place in a **nest box** inside the chicken coop. Sufficient straw should be provided to help maintain the correct temperature.

Artificial incubation

Large amounts of eggs are often incubated artificially. In general, incubators will let you control the **temperature** and **humidity** inside the incubator. Some incubators rotate the eggs.

Eggs should be collected for incubation as soon as possible and cleaned with multipurpose virucidal **disinfectant** before putting in the incubator.

The amount and frequency of **egg turning** needed differs for each species, but each egg needs turning at least twice per day.

Eggs in an incubator.

Egg candling

Eggs should be checked at 10 days to ensure that incubation is progressing. This is done using an **egg candler**, a light used to illuminate the egg so its development can be assessed. Under the candler:

- viable eggs will show a network of blood vessels
- eggs that contain a dead chick will show a circle of blood vessels
- eggs that were never fertilised will lack any indication of blood vessels.

Any eggs that are not viable should be disposed of.

Incubation

Incubation periods and temperatures for commonly bred poultry are shown in the table.

Species	Incubation period (days)	Incubation temperature (°C)
Chicken	21	37.5
Pheasant	24–26	37.5
Duck	28	37.5
Goose	28–33	37.5
Turkey	28	37.5
Ostrich	42	36.4

Hatching

If using a large multi-tray incubator, eggs due to hatch imminently should be moved to a hatcher. Only interfere with the hatching process by keeping the membranes moist if they become stuck on the chick or if the chick becomes exhausted. Chicks should remain in the incubator until they are dry and fluffy.

Records

Eggs should be marked for identification and detailed records kept of parentage, if known. Also include the date they were laid, initiation of incubation and observations during regular candling. Chicks should be weighed once fluffy.

Now try this

Describe a hygiene plan for an egg incubation facility.

Care plans: poultry – hatching to fully grown

Housing and treatment of a sick chick will differ greatly between backyard breeders and commercial producers, and whether they are kept for egg production or meat production.

Housing

Once fluffy, chicks can be removed from the incubator and maintained in a pen (brooder). A **heat lamp** must provide a temperature underneath of 30°C, with space for the chicks to move away from it to control their own temperature.

An **absorbent substrate**, such as pine shavings, corn cobs or rice hulls, should be used to prevent sore feet, which can lead to bumblefoot.

Young chicks are especially at risk of drowning, so it is important to use only **shallow water dishes**.

Enrichment can both reduce harmful behaviours and increase productivity.

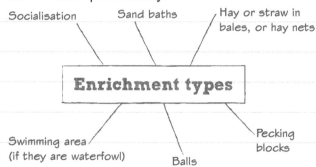

Socialisation   Sand baths   Hay or straw in bales, or hay nets

Enrichment types

Swimming area (if they are waterfowl)   Balls   Pecking blocks

Nutrition for chickens

Chicks should initially be fed **ad-lib** to encourage **foraging** behaviour. Once they are more established, the amount of food given can be reduced. A starter crumb is given until about six weeks, followed by complete starter feed and grit. Layers will need to proceed on to a high protein complete layer feed, plus grit and oyster shells.

Healthcare

Worming: Hairworms, roundworms and gizzard worms are common **intestinal parasites**. Poultry should be **wormed** every 3–6 months. Any eggs laid within at least seven days of worming should not be eaten.

Vaccinations: It is recommended that poultry are given a blood test prior to vaccination to determine if the vaccine is needed. Possible **vaccines** include:

- infectious bronchitis
- avian rhinotracheitis
- mycoplasma gallisepticum
- salmonella
- Marek's disease.

Red mite

An ectoparasite that feeds off the animal's blood leading to anaemia and reduced productivity. Treatments include mite powder and thorough cleaning of the coop.

Scaly leg mite

An ectoparasite that burrows under the scales in the legs. Treatments include anti-parasites and covering the legs in petroleum jelly.

Monitoring and recording keeping

Chicks should be weighed regularly. Any chicks failing to thrive should be given a full health check to identify any underlying causes.

Common problems

Crop impaction

A blockage in the crop linked to an unsuitable diet or the presence of unsuitable materials in the animal's environment. Fluid therapy can be used, but surgery may be required to remove the blockage.

Feather pecking

This is caused by a range of conditions including stress and unsuitable conditions.

Now try this

Describe how pecking issues can be limited in a large poultry unit.

Care plans: reptiles

Before breeding reptiles, it is essential to check if the species is egg laying or live-bearing. This will have a big impact on how the animals are cared for.

Breeding

For successful breeding of reptiles that hibernate, their environmental temperature and light levels may need to be lowered for 2–3 months and then increased prior to introducing the males and females. Some reptiles should be fed increased amounts prior to breeding and until egg-laying.

Egg laying

The female should be provided with a secure area (**deposition site**) to lay her eggs. Often a plastic tub with a hole in the lid containing a moist substrate can be used. The substrate will vary depending on the species, for example turtles need over eight inches depth of soil to lay their eggs in.

Incubation

Most reptile eggs require **artificial incubation** to regulate and maintain the temperature and humidity levels. Perlite or vermiculite substrates retain moisture and allow airflow. The **incubation period** depends on the temperature and humidity. Generally, lower temperatures cause longer incubation times.

Eggs should be transferred to the incubator within a few days of laying. If they were laid in a clump or attached to an object, they should not be separated in case of damage.

The eggs of some species develop better in the dark. For some reptile species, incubation temperature controls the **gender** of the hatchlings.

Species	Incubation period
Bearded dragon	50–80 days
Corn snake	58–62 days
Tortoise	8–11 weeks, but can take over 18 weeks
Red eared slider turtle	50–64 days

Good eggs v bad eggs

Good eggs usually have a consistent creamy shell, while bad eggs often look yellowish. However, external appearance does not guarantee success: some good eggs are infertile, while some bad eggs will develop and hatch well. All eggs should be put in the incubator until you are sure they will not hatch. If the appearance of the eggs deteriorates during incubation, there may be a problem with the conditions.

Depending on the species, some eggs can be candled from 3–5 days after laying. However, reptile eggs should **not** be turned as turning and handling often lead to death of the hatchling. This is because, within a few days of laying, the embryo settles at the bottom of the egg, the air sac at the top.

Hatching snake eggs

Shortly before hatching, eggs often dimple and start to collapse. The hatchling will then make a small hole in the egg (known as **pipping**). Some animals will emerge from a pipped egg within hours, others may remain in the egg for days.

Neonatal care

Hatchlings are best kept separately in order to monitor health, food intake and prevent bullying between hatchlings. It is especially important for cannibalistic species like king snakes.

Just before hatching, most hatchlings absorb the remains of the yolk sac so they may not eat for a few days. Food items offered should be appropriate for the species and a suitable size. Generally, food offered to snakes should be no wider than the snake's belly while food offered to lizards should be no bigger than the space between the lizard's eyes.

Now try this

A corn snake has laid a clump of eggs. Explain why these should not be separated.

Congenital and hereditary conditions

There are many **congenital defects** in animals. You need to know some of the more common problems and be aware of how they can impact on the health and welfare of the animals.

Atresia	A passage in the animal's body is unusually narrow or absent. It can affect many parts of the body and requires veterinary advice. <ul><li>**Oesophageal atresia** – oesophagus ends in a pouch and so does not allow access to the stomach.</li><li>**Pulmonary atresia** – pulmonary valve does not form properly so blood cannot flow through the heart.</li></ul>
Cryptorchidism	One or both of the **testes** do not descend from the abdominal cavity to the scrotum, leading to **fertility problems**.
Glycogen storage disease	The animal does not have all the enzymes needed to convert **glycogen** (energy storage molecule in the body) into the form of energy the body can use, resulting in poor growth. Observed in: dogs, cats, cattle, quail, chickens, mice and rats.
Hernia	An abnormal opening in muscle, allowing other body tissues to pass through. There are three main types of hernia: **inguinal** hernia, **umbilical** hernia and **diaphragmatic** hernia. If the protruding tissue becomes strangulated (has its blood supply cut off), there is a risk of infection and tissue death.
Poor weight gain	General **poor development** and low weight gain.
Polydactyly	The animal has **extra digits** on one or more limbs. In some species, this is encouraged, such as dogs working on snowy ground where extra grip is beneficial.
Scoliosis	Crookedly aligned vertebrae, leaving the animal with a **curve** in their spine.

Breed profile of Norwegian Lundehund

Forepaws must have at least six toes and eight pads. Five toes must rest on the ground. Hind paws must have at least six toes, four of which must rest on the ground.

A Norwegian Lundehund paw.

Congenital vs hereditary

A **congenital** condition is one present from birth. A **hereditary** condition is one that is inherited from the parents.

A cocker spaniel puppy with a hernia.

Now try this

Polydactyly is an inherited condition.

Explain the impact this can have on health and welfare in dogs.

Analysing DNA

Modern technology allows us to study the **genome** of an organism. **DNA** can be extracted from cells and amplified so that it can then be sequenced and genes identified. Organisms for which genomes have been sequenced include nematodes, mosquitos, fruit flies, puffer fish and humans.

DNA extraction

- Cells in the sample are broken open by blending or grinding.
- Fats and proteins in the mixture are emulsified by adding detergent and salt.
- Ethanol is added and the sample centrifuged to separate DNA from the rest of the cell debris.
- The extracted DNA can then be amplified, sequenced and analysed.

Restriction enzymes are used to cut DNA into shorter chains or fragments. They are specific, so only cut DNA at a particular sequence, leaving sections of unmatched DNA on each strand called **sticky ends**.

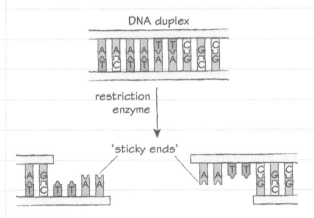

DNA duplex

restriction enzyme

'sticky ends'

Polymerase chain reaction (PCR)

A PCR machine is used to amplify (make lots of copies of) the extracted DNA fragments, which can then be used for gene isolation, sequencing, cloning, mapping mutations or studying gene expression.

- Samples are put into the machine along with DNA polymerase, primers and nucleotides.
- A cycle of automated temperature changes starts the synthesis of new DNA strands by adding complementary nucleotides onto the primer.
- Each repeat doubles the number of DNA fragments.

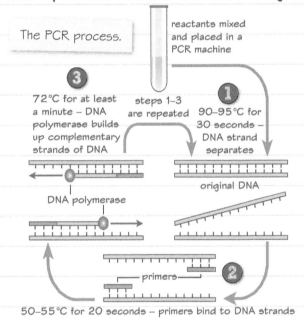

The PCR process.

reactants mixed and placed in a PCR machine

③ 72°C for at least a minute – DNA polymerase builds up complementary strands of DNA

steps 1–3 are repeated

① 90–95°C for 30 seconds – DNA strand separates

DNA polymerase

original DNA

primers

② 50–55°C for 20 seconds – primers bind to DNA strands

DNA profiling

This technique can identify individuals or relationships. It uses the fact that restriction enzymes cut an individual's DNA into fragments in a way that is unique to the individual. These fragments are separated by **electrophoresis**. This gives a pattern that can be used to compare the relationship between two different animals, such as to check that an animal's pedigree is accurate.

Gel electrophoresis

Electrophoresis can be used in DNA profiling or to identify a gene or analyse proteins. Different sections of DNA travel different distances depending on how long they are. The DNA is seen as bands which can be identified using a dye, UV fluorescence or DNA probes specific to a particular gene. Different samples will be analysed at the same time. Samples which have bands that travel the same distance contain the same gene.

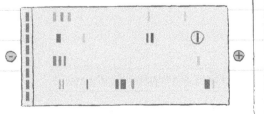

Now try this

A Dalmatian breeder has been told that one of the dogs he uses for breeding has been diagnosed with ciliary dyskinesia. He wants to arrange for the puppies bred from that dog to be genetically tested for the disease.

Explain how the animals' DNA would be analysed.

Genetic screening and disorders

Genetic screening is used to assess the genes of an animal. This can identify potential problems in offspring, or whether or not an animal is a carrier for a disorder. If animals in a breeding programme are suspected of having a genetic disorder, they might be screened to help decide if they are suitable for the breeding programme.

Screening for disorders

DNA can be **analysed** to identify parentage, breeding coefficients and the presence of recessive alleles. It can also be used to look at individual genes to see if the animal has a characteristic that could be desirable for **selective breeding**, or whether the animal or its offspring could suffer from **genetic disorders**. This could be used in combination with IVF and embryo transfer to select desirable gametes or embryos.

Genetic testing

This can be carried out on blood, skin, amniotic fluid or fur. It can be used for:

- ✓ parent identification
- ✓ disease screening (inherited)
- ✓ genotype profiling
- ✓ breed identification.

Severe Combined Immunodeficiency

Severe Combined Immunodeficiency (**SCID**) occurs in humans, dogs, mice and Arabian horses. It is an **autosomal recessive disease** (both parents must pass on a copy of the recessive gene for the offspring to have the disease). The animal has a malfunctioning immune system, leading to death within a few months. A high proportion of Arabian horses are carriers, so genetic testing is used in breeding programmes.

Von Willebrand disease

This is an autosomal recessive blood clotting disorder that can occur in humans and dogs. It leads to difficulties controlling bleeding after an injury, and is associated with spontaneous bleeding.

Freemartinism

Freemartinism occurs in cattle where a male and female twin develop simultaneously. The foetuses share the placenta, where hormones from both mix. The female twin may be infertile, show characteristics of a male reproductive system and have some cells with XY chromosomes. Genetic testing can be used to confirm freemartinism in calves to aid the early detection of a sterile female.

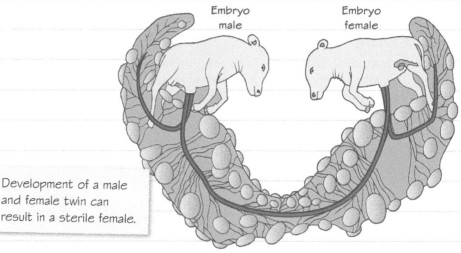

Embryo male   Embryo female

Development of a male and female twin can result in a sterile female.

Now try this

One puppy in a recent litter has been diagnosed with Von Willebrand disease.

Discuss how gene testing can be used to reduce the occurrence of this in future litters.

Gene modification

A genome can be changed to improve the productivity and health of an organism, as well as allowing us to study diseases more fully and manufacture proteins for use in medicine. **Genetic engineering** is the use of technology to genetically modify organisms, it can be used to analyse the function of specific genes through gene regulation.

Gene therapy

In **gene therapy**, techniques are used to make changes to the organism's genome to try and correct genetic disorders. There are three main approaches to gene therapy:

1 inserting functioning genes where the animal's own genes are not working

2 inactivating genes that are not functioning correctly

3 introducing new genes.

Unless the changes are made to the sex cells, these genetic changes will not be inherited by offspring.
Somatic cell gene therapy targets the affected body cells.
Germline gene therapy acts on ova, sperm and early embryos.

'Knockout' mice

These are lab-bred mice that have a section of DNA (a gene) **inactivated**, often by inserting artificial DNA. They are used to investigate the effect of a particular gene in order to study genetic disease and develop treatments. However, sometimes genes have different effects in different species.

Transgenics

In transgenics, a foreign gene is incorporated into an organism's DNA (see page 35, creating a transgenic animal).

Gene pharming

A mix of transgenic techniques and farming can be used to develop animals that produce proteins of medicinal or commercial value. These proteins may be secreted in milk, or released through eggs, making them easy to harvest.

The term 'pharming' is from farming and pharmaceutical.

Transgenics in action

Insulin, used to treat Type I diabetes in humans, used to be obtained from cows. Recently, bacteria have been genetically modified to produce large amounts of insulin identical to that produced in humans. There is less chance of the body rejecting this insulin, and no animal welfare implications.

Gene pharming in action

Alpha-I antitrypsin is produced in healthy livers. A deficiency in humans leads to increased liver cirrhosis and lung damage.

Sheep have been genetically modified to produce alpha-I antitrypsin in their milk that can then be used for treating humans.

Cloning

In **cloning**, an exact copy of an organism is made. DNA is extracted from an adult animal and used to replace the DNA in a fertilised ovum. The embryo is then implanted in a surrogate mother. The offspring will be genetically identical to the adult that donated the DNA.

Dolly the sheep

The first successfully cloned mammal, born in 1996, was Dolly the sheep, cloned from a mammary cell of a 6-year-old ewe. Dolly was euthanised at the age of 6 due to arthritis and progressive lung disease.

Now try this

A company is trialling a technique which can be used to genetically engineer cows so that they produce alpha-1-antitrypsin in their milk.

Identify and describe the technique being used.

33

Genetic modification process

Genes can be extracted from one organism and added into another organism's DNA to make **recombinant DNA**. Several methods can be used to transfer the recombinant DNA into a cell.

Producing recombinant DNA

(1) Isolation of the gene – three different methods:

- **restriction endonuclease** can cut out a section of DNA including the gene
- **reverse transcriptase** can make DNA from mRNA
- sequence can be worked out from amino acid sequence and DNA built to match.

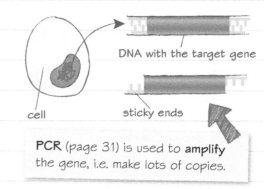

DNA with the target gene

cell   sticky ends

PCR (page 31) is used to **amplify** the gene, i.e. make lots of copies.

(2) Inserting gene into a vector for transfer:

- **plasmids** are often used to modify bacteria
- plasmid DNA is cut with same endonuclease as was used to isolate genes
- this leaves **complementary sticky ends**
- gene and open plasmid are mixed and bases allowed to pair
- **DNA ligase** is used to join sugar phosphate backbone.

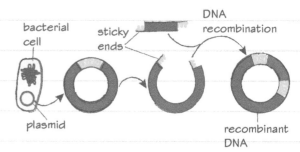

bacterial cell   sticky ends   DNA recombination

plasmid   recombinant DNA

(3) Getting recombinant DNA into cell:

- use of a virus
- DNA enclosed in **liposome**
- gene guns
- microinjection.

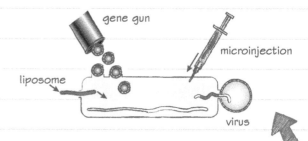

gene gun

microinjection

liposome

virus

Using a virus to transfer recombinant DNA into a cell is called **transduction**. Non-viral methods are called **transfection**.

Promotor

Some of the target genes may not need to be expressed in all cells, so it is important that a suitable **promotor** is used before adding the gene to the plasmid. The **promotor** is a section of DNA that, when it is active, instructs a cell to copy the section of DNA after it (telling the cell to activate the section of DNA).

Marker genes

Not all cells will successfully take up the recombinant DNA. To enable identification, **marker genes** are added. The marker gene has an easily recognisable phenotype (e.g. antibiotic resistance or colour change) that can help to identify which cells have taken up the gene.

The unit containing the recombinant DNA, promotor and marker gene is called a **construct**.

Now try this

In 1982, insulin from genetically modified bacteria was accepted for use in humans. Explain the process for producing bacteria which manufacture insulin. Explain the advantages and disadvantages of using insulin from microbes compared to insulin from animals.

Creating a transgenic animal

Animals that have had their DNA altered by the addition of a gene not normally found in that species are called **transgenic animals**. In animals, genetic modification usually occurs in embryos so that only a small number of cells have to be altered for the new gene to have an effect.

Transferring the recombinant DNA

- Transgenic animals are most commonly created by microinjection (**transfection**) of the **recombinant DNA construct** (page 34) into a fertilised egg.
- Alternatively, genes can be transferred to embryonic stem cells.

Gene integration

Some (but not all) of the eggs/cells will assimilate the new DNA with their original DNA. Any embryos that are made from these will have the modified DNA.

The DNA of the resulting offspring is tested using **PCR** technique to see which animals have the modified DNA.

Gene expression

In the cells where the **promoter gene** (page 34) is active, the cell will carry out the instructions of the modified DNA. This could be the production of proteins which might influence how the animal looks or how it functions.

Inheritance

Because the modified sections of DNA are present in the cells in the transgenic animal, it is possible for their gametes to contain the modified sections of DNA. When they reproduce, the modified sections of DNA may be passed on to their offspring. Because the inheritance of the modified DNA is not guaranteed, transgenic animals are often inbred for many generations to encourage the propagation of the modified DNA.

Creation of a transgenic cow

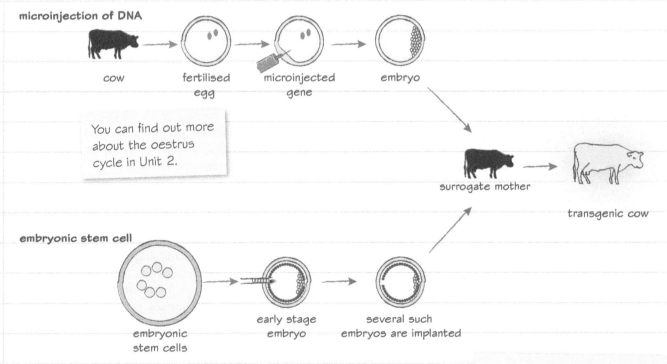

microinjection of DNA

cow → fertilised egg → microinjected gene → embryo

You can find out more about the oestrus cycle in Unit 2.

surrogate mother → transgenic cow

embryonic stem cell

embryonic stem cells → early stage embryo → several such embryos are implanted

This is an example of pharming.

Now try this

Transgenic sheep have been developed to secrete tissue plasminogen activator which dissolves blood clots in humans. Explain why the modified DNA is inserted into fertilised eggs instead of adult ewes.

Implications of genetic manipulation

Genetic manipulation can have a range of benefits and drawbacks, commercially and socially as well as practically.

Advantages of genetic manipulation

👍 Animals and plants can be altered to **increase** productivity, their nutritional value and disease resistance.

👍 Organisms can be modified so that they produce compounds of **therapeutic use** (pharming).

👍 Animals can be developed to model diseases, furthering **medical research**.

👍 Animals and plants could be bred to be **resilient to extreme conditions**, increasing hardiness and survival rates, and allow use of extra land.

👍 It is a **quicker** way to produce new breeds and increase variety.

👍 **Xenotransplantation** could be used to treat illnesses in a range of species.

👍 Animals can be bred to **remove characteristics** which have a high risk of injury, e.g. cattle bred without horns.

Disadvantages of genetic manipulation

👎 **Genetic variation** may be lost, resulting in lowered disease resistance and a lower ability to survive in a changing environment.

👎 Research and development is **long and costly**, with success not guaranteed.

👎 Facilities are not readily available for use with all animal collections.

👎 There are **restrictions** on when and how genetic modification can be used.

👎 There may be **health risks** due to the genetic modification, for example, in some breeds of pigs that have been bred to fatten quickly, the heart does not develop quickly enough to cope with the animal's body size.

👎 Potential transmission of **zoonoses** can be linked to the modified genes.

👎 **New diseases** may develop.

👎 There may be risks to animal health and the environment.

👎 It takes a long time to develop gene modification processes.

Polled cattle are bred not to have horns.

Case study

Research is being carried out into the development of cattle that are unable to form the infectious prions that cause bovine spongiform encephalopathy (mad cow disease).

Benefits could include: improved health for the cattle, less risk of illness to humans, less risk of loss of income due to illness in the herd.

Drawbacks could include: cost of research and gene editing, cost of modifying animals, cost of culling to promote selection of the new DNA in offspring, implications for health due to the modification, potential impact on the environment.

Now try this

High levels of phosphorus, found in fertiliser based on pig waste, have a negative impact on marine life. Enviropigs were developed to metabolise phosphorus, so less would be present in their waste.

Discuss the advantages and disadvantages of using Enviropigs.

 Most of the advantages and disadvantages given on this page will apply to different scenarios. Think about how you could use the points listed to make them specific to Enviropigs.

Regulation and ethics of genetic manipulation

The genetic modification of animals raises many **ethical concerns**. They can impact animal breeding practices.

Environmental risks – due to the creation of new and altered organisms which could impact on the biosphere, e.g. modified organisms created in captivity could eventually become part of the wider ecosystem

Religious and ethical implications for food animals – e.g. transgenic animals which contain elements of another species would be problematic for religions or ethical beliefs which restrict foods from particular animals

Links You can revise the advantages and disadvantages of genetic modification on page 36.

Animal rights – animals could become commodities, e.g. patented genetic sequences allow ownership of animals exhibiting this sequence

Ethics of genetic manipulation

Moral issues – the extremity of the research, its purpose and intent, e.g. GMO to reduce a food shortage compared to GMO to produce exotic species

Animal welfare – animals intentionally created with defects or disorders will have a lower welfare standard, e.g. mice which are genetically modified to develop cancer

Impact on non-modified organisms – due to the modification of other organisms, e.g. disease-resistant animals and plants could impact on other animal's ability to survive

Regulation and accessibility – determining regulations which should govern genetic modification and approval protocols, e.g. should genetically modified organism (GMO) products be controlled by the government?

Species boundaries – inserting genes from a species into one which would not naturally have those genes, e.g. Flavr Savr tomatoes

Regulation of the use of genetically modified organisms (GMOs)

In the UK, the use of GMOs is regulated by the Genetically Modified Organisms (Contained Use) Regulations 2014.

Featherless chickens (which don't require plucking) have been proposed as a time- and cost-saving measure. There are welfare concerns as feathers protect chickens from harsh weather and parasites.

Control of GMOs in food

GMO foods have to be approved and labelled appropriately by the European Food Safety Authority (EFSA) before they can become available for human consumption. Products from animals which have been reared on GM foods do not need identifying when they are for sale.

Common GM animal feeds include GM maize and GM soya. No GM crops are currently grown in Britain.

Now try this

Several companies have recently developed fast-growing salmon to make salmon farming more profitable.

1. Describe **one** advantage and **one** disadvantage of this approach.
2. Outline **three** possible ethical issues faced by salmon farmers using genetically modified salmon.

Your Unit 1 set task

Unit 1 will be assessed through a task, which will be set by Pearson. In this assessed task, you will need to answer questions based on your understanding of **animal breeding and genetics**.

Set task skills

Your assessed task could cover any of the essential content in the unit. You can revise the unit content in this Revision Guide. This skills section is designed to **revise skills** that might be needed in your assessed task. The section uses selected content and outcomes to provide examples of ways of applying your skills.

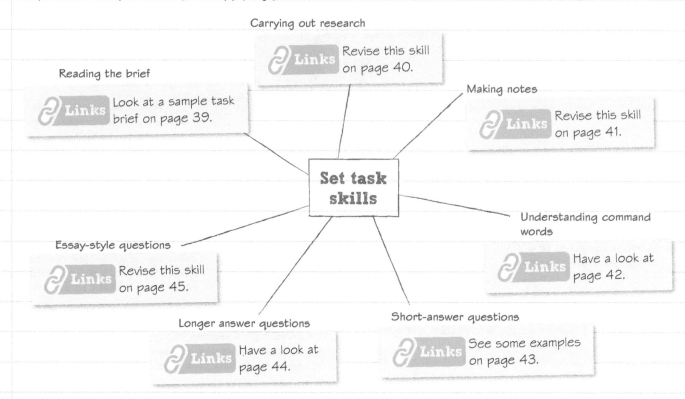

Carrying out research
Links Revise this skill on page 40.

Reading the brief
Links Look at a sample task brief on page 39.

Making notes
Links Revise this skill on page 41.

Set task skills

Understanding command words
Links Have a look at page 42.

Essay-style questions
Links Revise this skill on page 45.

Longer answer questions
Links Have a look at page 44.

Short-answer questions
Links See some examples on page 43.

Check the Pearson website

The activities and sample response extracts in this section are provided to help you to revise content and skills. Ask your tutor or check the Pearson website for the most up-to-date **Sample Assessment Material and Mark Scheme** to get an indication of the structure of your actual assessed task and what this requires of you. The details of the actual assessed task may change so always make sure you are up to date.

Now try this

Visit the Pearson website and find the page containing the course materials for BTEC National Animal Management. Look at the latest Unit 1 Sample Assessment Material for an indication of:

- the structure of your set task, and whether it is divided into parts
- how much time you are allowed for the task, or different parts of the task
- what briefing or stimulus material might be provided to you
- any notes you might have to make and whether you are allowed to take selected notes into your supervised assessment
- the questions you are required to answer and how to format your responses.

Reading the brief

Here are some examples of skills involved if reading a brief and planning which topics you need to research.

Read the task information carefully. Highlighting the key points will help you clearly see what types of information you need.

Practising your skills

This revision task brief is used as an example to show the skills you need. The content of a task will be different each year and the format may be different. Check with your tutor or the up-to-date Sample Assessment Material on the Pearson website for details.

Revision task brief

'Herptilia' is a local reptile breeder and pet store. It provides a range of reptiles, amphibians and invertebrates to the public.

The manager of Herptilia has asked you to provide information to help the company develop an effective approach to breeding animals. The manager has specifically requested information about variations for bearded dragons:

- scaling types such as leatherback and silkback
- colour morphs.

Information the manager would like clarifying relates to husbandry needs, care requirements, dietary requirements, mating/reproductive behaviours and neonatal care.

Sample notes extract

You could mind map the key points you have identified to help you research each topic fully.

Scaling – normal, leatherback, silkback   Colours

Neonatal care: monitoring, stages of development, handling, information specific to this life stage

Bearded dragon

What information do I need to find?

Desirable features for breeding

Your mind map does not have to form part of your notes, but could help you structure your research and notes.

Mating/ reproductive behaviours

Using different colours could help you find information on different topics later.

Care requirements: health checking, preventative treatments, routine treatments, handling and approaching

Dietary needs: including suitable and unsuitable feeds, supplements, common dietary problems, changes with physiological state

Husbandry requirements: cleaning, bedding, space

It is a good idea to spend some time writing the plan for your research and notes.

Sample notes extract

What general information should I cover?

- Genetics: monohybrid and dihybrid crosses, phenotypes
- Deciding which animals to breed: temperament, phenotype, genotype
- Identification of oestrous/pregnancy
- Gene manipulation techniques
- Ethics

Remember you can include general information in your notes too and use simple bullet points.

 Links To revise principles of breeding management, refer to pages 9–12.

Now try this

Read the revision task brief at the top of this page, and draw your own mind map or set of bullet points to help you plan the research you would carry out.

Carrying out research

After you have identified the key areas to research, you need to think about where you can find useful information. You may need to spend more time researching the topics you don't understand so well.

Where to look for information

✓ Books

✓ Periodicals: newspapers and magazines

✓ Journals

✓ Reports: especially those published by DEFRA or animal organisations (e.g. RSPCA)

✓ Online libraries: your centre may have a subscription to an online library

✓ Internet: especially breed society web pages (e.g. the Kennel Club or the Governing Council of the Cat Fancy)

✓ Google Scholar: allows you to search online academic articles.

Identifying good research sources

Make sure your sources are:

1 **Relevant:** Check that the source gives you the information you need for your set task. Don't include information just because it looks detailed.

2 **Current:** Science evolves all the time. Information from 10 years ago may already be outdated.

3 **Reliable:** Not all information on the Internet is accurate or trustworthy. Consider whether each website you look at contains accurate information.

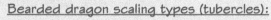

These research notes are based on the revision task brief on page 39 and include key information on scaling types and colours. You should check the latest Sample Assessment Materials to see what sort of information you will be given in your assessment.

Sample notes extract

Bearded dragon scaling types (tubercles):

Normal: larger scales, spikes present. Found naturally. Homozygous dominant.

Leatherback: smaller scales, few spikes. American and Italian breeds. Heterozygous.

Silkback: no scales, no spikes. Produced by breeding two leatherbacks. Homozygous recessive.

Colours:

Brown/tan: most common in the pet trade.

Yellow/citrus: common in a range of shades from bright yellow to pale gold.

Red or orange: highly collectable due to rarity

White: can be albino or a very pale colour.

Translucent: patches of skin with no colour, black eyes; these lack the ability to make white pigments. Translucence decreases with age as skin thickens. Can have poor health and shorter life span due to inbreeding.

Hypomelanistic: do not produce melanin, which is the pigment responsible for darker markings (common in wild bearded dragons).

Including phenotype information and using correct technical terms will help you to show accurate and thorough understanding of the topic.

Include as much **relevant** detail in your notes as you have space for.

Links To revise desirable characteristics in herptiles, see page 8.

Make sure you plan your research time carefully.

Make a note of any useful websites you find that provide good information.

Now try this

Research the mating and reproductive strategies of bearded dragons and prepare some rough notes.

Try setting yourself a time limit for this research. See how much you can find out in 10 minutes.

 Links To revise breeding and reproductive strategies, see pages 17 and 18.

Making notes

Here are some examples of skills involved if making notes.

These notes are based on the revision task brief on page 39. They are adequate and include relevant information about enclosure sizes, temperatures and atmosphere. They differentiate between conditions for adults and young.

Sample notes extract

Bearded dragon husbandry needs

Enclosure size: min 90 × 45 × 45cm. Ideal 120 × 60 × 60cm

Enclosure type: wooden or glass vivarium

Lighting: UVB and heat lamp. Regular day and night cycles.

Temperature: 26–32°C at the cool end. Basking end: 35–40°C for a neonate, 35°C for a juvenile, 33°C for an adult (approx.).

Humidity: low (35–40%)

Substrates: sand, corn cob, paper, reptile carpet

Preparatory notes

You may be allowed to take some of your preparatory notes into your supervised assessment time. If so, there may be restrictions on the length and type of notes that are allowed. Check with your tutor or look at the most up-to-date Sample Assessment Material on the Pearson website for information.

Improved notes extract

Bearded dragon husbandry needs

Enclosure size: min 90 × 45 × 45 cm. Ideal 120 × 60 × 60cm; more height can cause problems establishing the temperatures needed.

Enclosure type: wooden or glass vivarium with suitable ventilation for temperature, humidity and air circulation.

Lighting: UVB and heat lamp near to each other for calcium production. Heat lamp should be at one end of the enclosure and near the UV light. Regular day and night cycles, min. 12 hours UV light per day.

Humidity: low: 35–40% to mimic natural desert environment.

Temperature: Ideal temperature range of 26–32°C

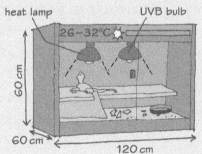

heat lamp   UVB bulb
26–32°C
60 cm
60 cm
120 cm
35–40°C (neonate)
35°C (juvenile)
33°C (adult)

Substrates: sand (natural but can cause impaction), corn cob (cheap but can cause impaction), paper (cheap but poor absorbency), reptile carpet (reusable, not natural).

Try to make your notes as detailed as time and space allows. These improved notes include more details about the correct environment, and include reasons, pros and cons, and price comparisons.

Your notes can include diagrams if you find them useful. These can be a good way of presenting information succinctly, but remember they only need to show essential detail so don't spend too long on them.

The notes must be useful **for you**. Don't make pages of notes with information you already know well, if this uses valuable space that you need for notes on topics you are unsure of.

 Links To revise husbandry requirements, see page 24.

Now try this

Prepare some brief notes about the ethical implications of breeding bearded dragons.

 You will need to select appropriate sources and create condensed but informative notes. Are there any specific examples you could include?

Understanding command words

It is important that you understand what skills are expected of you in your assessment. Here is some guidance on what skills are expected for the different command words used within the questions.

Worked example

Each question contains a **command word**, which tells you what sort of response is expected. Read the question carefully and use the marks available as a guide to how much information to include. These examples show how different command words require different responses.

1 **Give** one example of a use for the bottle shown. `1 mark`
To provide drinking water for animals

2 **Describe** the item shown in the picture. `2 marks`
It is a blue plastic water bottle, with a metal drip nozzle, which holds approximately 320 ml of liquid.

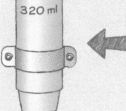
320 ml

This is a good **description** with plenty of detail.

3 **Explain** what the item shown is used for. `2 marks`
It is a water bottle; its purpose is to contain liquids and allow an animal to drink from it.

This answer is incomplete. You would go on to give further positive features and then the negatives.

4 **Evaluate** the fitness for purpose of the item shown as a water bottle for animals. `6 marks`
It is fit for purpose due to the water-resistant material it is made of. It has the potential to be reused, making it good value and reducing its impact on the environment …

Check the marks available. Higher mark questions require more in-depth answers. This answer is incomplete. A full answer would explore the suitability of other methods for different species.

5 **Discuss** the features of different methods of providing water for a variety of animals. `8 marks`
Water bottles for animals are usually made out of plastic, while water dishes can be made of plastic, metal or ceramic. Both bottles and dishes come in a variety of sizes, making them suitable for a range of animals. The method selected will depend on the animal it is for. Small mammals, such as rodents, rabbits and guinea pigs, naturally drink from ground level so a dish would serve this purpose well …

Other command words

Other command words you could encounter in your assessment include:

- **calculate** – give a numerical answer, and show your working
- **define** – give the meaning of a term
- **state** – give clear examples or facts about the subject
- **assess/analyse** – consider different factors of the topic and identify those that are most important or relevant; you may be asked to make a recommendation, giving your reasons.

Planning your response

Make sure you think before you write:

1. What am I being asked to do? Look at the command word.

2. What do I know about the topic?

3. How much information do I need to include? Look at the number of marks available.

4. Do I need to structure my answer and, if so, how?

5. For longer answer questions, consider making a list or spider diagram of points you want to include.

Read through your answer. Does it answer the question you've been asked?

Now try this

1 Try completing questions 4 and 5 in the response above.

2 Evaluate the use of gene manipulation techniques in the development of disease-resistant animals.

For an 'Evaluate' question, include enough points to produce a balanced argument.

 Links To revise gene manipulation techniques, see page 33.

Short-answer questions

Some questions will only require short answers. The types of command words often used in this type of question are: 'state', 'give', 'describe', 'define' and 'explain'. They will typically be worth between 1–4 marks.

Worked example

1 Scale types in bearded dragons are thought to be the result of incomplete dominance.

a) Explain the type of gene interaction this shows. **2 marks**

Sample response extract

Incomplete dominance is a gene interaction where the heterozygous animal shows an intermediate phenotype to the homozygous phenotypes. Neither allele is dominant so a mid-point between both homozygous phenotypes is expressed.

Ask yourself: 'What information could I give?' Here the relevant information to include is:

Neither allele is dominant.

The heterozygous phenotype is an intermediate phenotype.

b) Explain how this relates to homozygous and heterozygous phenotypes. **2 marks**

Sample response extract

In bearded dragons, the homozygous dominant phenotype (normal) is for rough scales, while the homozygous recessive phenotype (silkback) has no scales. Small scales in leatherbacks are the intermediate phenotype seen in heterozygous animals.

Link your information to the species you are discussing.

 Links To revise gene interactions and phenotypes, see pages 2–5.

Tips for short-answer questions

✓ Read the question carefully. If the information you have written does not answer the question, you won't get any marks for it, even if it is accurate.

✓ You won't usually need to plan your answers to short-answer questions, but you may want to jot down a key word or two first.

✓ Don't write too much. Spending too long on short-answer questions may leave you short of time for questions worth more marks.

✓ Give an answer to every question, even if you're not 100% sure it's right.

✓ Remember to check that your answer makes sense.

Now try this

Give **two** innate mating behaviours seen in rabbits.

Longer answer questions

Command words such as 'discuss', 'explain', 'assess', 'analyse' and 'evaluate' require you to go beyond just recalling facts. These questions will typically be worth between 6–12 marks.

Worked example

1 Discuss **three** factors which should be considered when selecting animals to breed for the 'Herptilia' pet shop.   **6 marks**

Sample response extract

Animals should be mature enough, but not too old, to breed without damage to their health. The animals should be healthy. The animals should have a good temperament.

> This would be an acceptable response to a question such as 'state three factors …', but it is much too brief here. You must explain **why** it is important to consider each factor and what the **consequences** of ignoring these factors might be.

Improved response extract

To ensure the success of the breeding programme, it is important to select animals with care and consider a range of factors.

Firstly, animals should be mature enough, but not too old, to breed without damage to their health; this will help maintain the original stock and increase the chances of healthy offspring. Secondly, animals should be healthy and of a line with no inherited disorders. This is in order to limit the potential for congenital defects, encourage the health of any progeny and ensure the success of mating. Thirdly, consider the temperament of the animals. Naturally aggressive animals are less likely to mate successfully and more likely to produce aggressive offspring, which will make handling more difficult and reduce the number of potential buyers.

> An introduction to longer answers can help demonstrate that you understand the relevance of what you are being asked.

> This answer gives three factors, as required by the question. Giving more would be unnecessary and take up valuable time.

> **Links** To revise the factors to consider for breeding, see page 10.

Worked example

2 Explain, with an example, the impact of inbreeding on the health of bearded dragons.   **4 marks**

Sample response extract

Inbreeding bearded dragons can impact on the health of the offspring in a number of ways. It can lead to an inbreeding depression which will reduce the biological fitness of the offspring, and thus their chances of survival and reproduction. It can also reduce the offspring's resistance to disease. It will increase the chances of any detrimental genes in the animals' DNA being expressed, as the genetic variation reduces with every generation. An example of this can be seen in translucent bearded dragons, which have a shorter lifespan and are prone to illnesses.

> When carrying out your research, make a note of key examples you could include for the different topics, such as genetic variations. These could help you provide examples specific to the species of animal you are being asked about.

> **Links** To revise inbreeding, see page 7.

Now try this

Explain how Gregor Mendel's law of independent assortment is applied in dihybrid crosses.

> Make sure you know exactly what the question requires. Here, you first need to explain what the law of independent assortment is, before you then explain the combinations of alleles which can occur within the parent's gametes and in the offspring.

 Links To revise Mendel's laws, see page 3.

Essay-style questions

Some questions may require you to produce an essay-style answer. Your answer will be marked on both content and how well you have structured it.

Essay-style questions are designed to demonstrate your accurate, thorough knowledge and understanding of a given topic. Make sure that your answer:

- ✓ has a clear and logical structure
- ✓ is well supported by links to relevant evidence on the topic
- ✓ is clear and concise
- ✓ includes logical discussion throughout to support your points
- ✓ includes relevant links to the context given in the question
- ✓ if you include any diagrams, make sure they are well-developed and convey all the necessary information.

Worked example

1 Discuss the issues concerning the use of genetic modification in the breeding of animals. **12 marks**

Plan your answer to this type of question first, so you know what you want to include and can put the information in the most logical order.

Sample response extract

There are a range of potential benefits and drawbacks to using genetic modification in the breeding of animals.

An advantage of genetic modification is the potential to reduce or remove detrimental gene sequences from animals and replace them with beneficial ones, which could reduce the occurrence of hereditary disorders in animals.

However, genetic modification can also be used to induce disorders, as with the onco-mouse, which will negatively impact on the welfare of the animals.

A further advantage of the use of genetic modification in breeding animals is the potential to produce disease-resistant animals which would not only strengthen the breed as a whole over time, but allow for healthier young.

However, the presence of these animals could then have a negative impact on both non-modified organisms' survival and other animals in the food web ...

Remember, if you are asked to discuss something, consider different aspects of a topic and the links between them. An introduction and a conclusion can help structure your answer.

Notice here that both advantages and disadvantages are discussed.

The response might continue by outlining other concerns such as animal rigths and public safety as well as both the pros and cons of aspects of animal welfare that could be affected by genetic modification.

Links To revise genetic modification, see page 33.

Now try this

The answer to the example question above is incomplete. Make a plan for your own answer to this question.

For extra practice, have a go at actually answering the question in full.

Links To revise the moral, religious and ethical issues related to the use of genetic modification in breeding animals, see page 36.

Cellular ultrastructure

A **cell** is the most basic structural unit for all living organisms. It is the building block of life. All living things are made up of one of two types of cell: **eukaryotic** or **prokaryotic**.

Ultrastructure of a eukaryotic cell

plasma membrane
cilia
cytoplasm
cytoskeleton
centriole
lysosomes
peroxisome

mitochondria
endoplasmic reticulum (ER)
Golgi apparatus
nucleus
nucleolus
nuclear envelope
ribosome

Ultrastructure of a prokaryotic cell

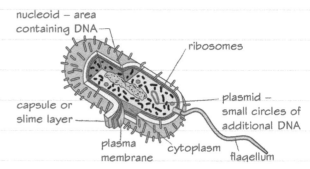

nucleoid – area containing DNA
ribosomes
capsule or slime layer
plasmid – small circles of additional DNA
plasma membrane
cytoplasm
flagellum

Eukaryotic cells have a **nucleus** and other organelles that are surrounded by a **membrane**. They are found in multicellular organisms.

Prokaryotic cells form **single-celled** organisms such as bacteria. They do **not** contain a 'membrane bound' nucleus or organelles.

Functions of organelles

Structure	Function
Nucleus	The cell's **control centre**. It coordinates all the cell's activities such as growth and reproduction (cell division), and stores the cell's DNA.
Nucleolus	Stores the cell's **RNA** and is important in the production of **ribosomes**.
Mitochondria	The site of **respiration** using oxygen. This produces ATP, the energy source for many life processes.
rER	Rough endoplasmic reticulum (rER) contains many **ribosomes** and is important in the formation and storage of **proteins**.
sER	Smooth endoplasmic reticulum (sER) is important in the production and storage of **lipids** (fats) and **steroids**.
Golgi apparatus	Works together with the ER, and is important in storing and packing molecules such as proteins and lipids (e.g. hormones and enzymes) in vesicles ready for transport.
Lysosomes	Contain **enzymes** whose main role is digesting waste products and debris. Some type of cells have **conventional lysosomes** that contain secretory products and are known as **secretory lysosomes**.
Plasma membrane	**Encloses** the cell's contents.
Cilia	Short microscopic hair-like structures. **Move fluid** past the cell's surface, and help the cell to move.
Cytoskeleton	Important in maintaining the cell's **shape**. The cytoskeleton is made of **microtubules** (position organelles), **actin filaments** (cytokinesis and cell movement), and **intermediate filaments** (provide strength and support).
Ribosomes	Made of two sub-units which join together to **manufacture proteins**. 80s ribosomes are found in eukaryotic cells, 70s ribosomes in prokaryotic cells.
Centrioles	Two hollow cylinders arranged at right angles to each other to form the **centrosomes**. They are important in **spindle formation** during cell division.
Peroxisomes	Small **vesicles** containing oxidative enzymes. Help to remove toxic substances.

Now try this

1　Where is all the genetic information stored in an organism?

2　Explain the function of mitochondria within the cell.

Links　See page 47 for more information on the plasma membrane.

The plasma membrane

The **plasma membrane** is made up of many different molecules. Each molecule has a different role. The fluid mosaic model is a model of the plasma membrane.

The fluid mosaic model

The main components of the plasma membrane are **phospholipids**. Phospholipids have a **hydrophilic** head that is attracted to water, and a **hydrophobic** tail that repels water. Phospholipids form a **bilayer**.

The bilayer contains many proteins. Some are **integral** and cross the membrane, and some are **peripheral** and are found only on one side of the membrane.

Some components of the plasma membrane are attached to long carbohydrate chains. Phospholipids attached to chains are called **glycolipids**. Proteins attached to chains are called **glycoproteins**. The plasma membrane also contains **cholesterol**.

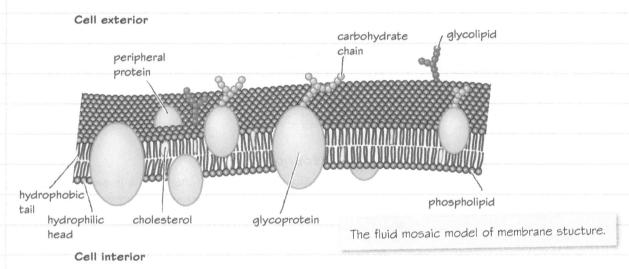

The fluid mosaic model of membrane stucture.

Role of the plasma membrane

The plasma membrane around the cell:

- is partially permeable, and so controls the transport of molecules in and out of the cells
- separates the contents of the cell from the outside environment
- allows other cells to recognise the cell as belonging to the body
- allows cells to signal to each other
- allows molecules, such as hormones or drugs, to bind to the receptors in the membrane
- holds the components of some chemical reactions, for example enzymes, in place.

There are also internal plasma membranes around the organelles, which separate the contents of the organelles from the cytoplasm and allow molecules to be transported in and out of the organelle.

Functions of the components of the plasma membrane

- ✓ Phospholipids – control what enters and exits the cell.
- ✓ Integral protein – transport of molecules in and out of the cell.
- ✓ Glycoproteins and glycolipids – cell recognition, cell signalling and receptors.
- ✓ Peripheral protein – enzymes.
- ✓ Cholesterol – maintains stability of the cell.

Now try this

Explain the role of plasma membrane around the mitochondria.

Remember it is made up of lipids and proteins to carry out specific functions.

Ensure your answer clearly states what the plasma membrane does.

Microscopy

In order to see cells and their structures, we need to **magnify** them. We can use **light microscopes** or **electron microscopes** (EM).

The light microscope

A light microscope uses **light** and different lenses to magnify objects to allow the image to be seen through an eyepiece. The best possible magnification, with good resolution, is about ×1000.

You do not need to remember the names of the parts of the microscope, but you do need to know how the light microscope is used and the types of images you can see with it.

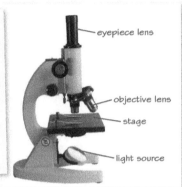

eyepiece lens

objective lens

stage

light source

The electron microscope

Electron microscopes use **electrons** instead of light to produce detailed images with a high magnification (×200 000 to ×2 000 000) and a high resolution. There are two types: **transmission** electron microscopes (**TEM**) and **scanning** electron microscopes (**SEM**).

How to work out the magnification

Photos of specimens taken down the microscope are called micrographs. It is easy to work out the magnification of the image if you know:

- the size of the image
- the actual size of the specimen.

The formula for this is:

$$\text{magnification} = \frac{\text{image size}}{\text{actual size}}$$

How to prepare a slide for a light microscope

If you are preparing a slide for a specimen that is in solution, for example, cheek cells in saliva, add a drop of specimen onto the slide and then add a drop of stain. You can then cover your specimen with a cover slip.

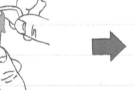

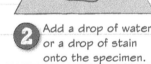

cover slip

slowly lower

water droplet

slide

1 Take a thin slice of your specimen and place it onto a clean microscope slide.

2 Add a drop of water or a drop of stain onto the specimen.

3 Take a clean cover slip and lower it slowly onto the specimen, taking care to avoid air bubbles.

The difference between magnification and resolution

Magnification is the number of times greater the image is than the specimen.

Resolution is the ability to distinguish between two points on an image. The higher the resolution, the sharper the image.

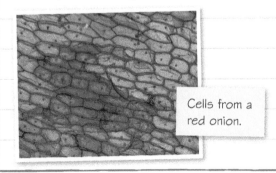

Cells from a red onion.

Staining a sample

Staining: used to see cells and cell structures more clearly. Different types of stains react with and colour different parts of the cell.

Gram staining: stains Gram positive bacteria red and Gram negative bacteria violet.

Eosin: stains red blood cells and cell membranes a pink or red colour.

Iodine: stains starch blue–black.

Methylene blue: stains nuclei and chromosomes blue.

Now try this

An image of an animal skin cell measures 15 mm across. The actual size of the cell is 0.0075 mm.

What is the magnification of the image?

Cellular control

In the mammalian body, there are vast amounts of **DNA** stored in almost every animal cell. DNA is the genetic information which is the blueprint for life.

Chromosomes and DNA

Chromosomes are found in the nucleus of cells and contain strands of DNA.

Chromosomes exist as pairs in all animal cells, except for egg cells and sperm cells. In a pair of chromosomes, there is one chromosome from the father and one from the mother. Each species has a different number of pairs of chromosomes. For example, dogs have 39 pairs of chromosomes.

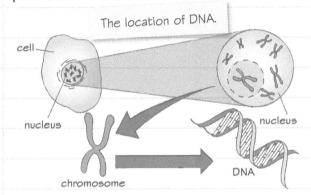

The location of DNA.

cell

nucleus

nucleus

chromosome   DNA

Cell division

Cell division has very important functions in **growth and repair** of body tissues and **sexual reproduction**.

Sperm and egg cells are produced my meiosis. Other cells are produced by mitosis. Mitosis produces exact genetic copies of the cell. The contents of the cell are doubled and then the cell divides into two. This results in **two genetically identical daughter cells** that are also identical to the parent cell. Each daughter cell is **diploid**, i.e. it has two complete sets of chromosomes.

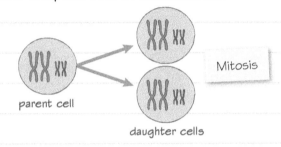

parent cell

Mitosis

daughter cells

The structure of DNA

DNA is formed by **nucleotides** and it carries the genetic information needed for cell reproduction, development and function.

DNA is made up of **bases** attached to a **sugar–phosphate backbone**. Each base contains nitrogen and is represented by a letter: **A, T, C or G**. These **pair** together and help to form the **double helix** shape.

A, adenine always **pairs with T**, thymine.

C, cytosine always **pairs with G**, guanine.

Genes

Genes are short sections of DNA that carry the genetic code for a particular characteristic or cell activity.

Within each gene, bases (A, T, C or G), form a structure called a **triplet**. Each triplet codes for a particular **amino acid**. The sequence of bases in a gene therefore codes for the sequence of amino acids in a **polypeptide chain**.

Polypeptide chains make up **proteins**. Proteins have a huge range of structures and functions in the living world, and they are all made of chains of amino acids.

Mutations in the genetic code

Mutations are caused by an **alteration** to the base pairing sequence in the genetic code.

There are different types of mutation. For example, a mutation that does not cause any change in the amino acid sequence in a protein is called a **neutral**, or **silent**, mutation, as it makes no difference to the protein.

Osteogenesis imperfecta

Alterations to the base pairing sequence as a result of mutations can cause problems such as **osteogenesis imperfecta** in dogs. This disease causes brittle bones, so dogs with this condition have an increased risk of injury and slow healing.

Now try this

Each parent cell divides by mitosis to produce two daughter cells during growth and repair of body tissues.
Domesticated horses have 32 pairs of chromosomes in each of their body cells.

Give the number of chromosome pairs each daughter cell will have.

Cell transport

Living cells of both plants and animals are surrounded by a **semi-permeable membrane** called the **plasma membrane**. This **regulates** the flow of liquids, and dissolved solids and gases, in and out of the cell.

Types of cell transport

Substances can be transported in and out of cells by different methods:

1 **active transport:**
- endocytosis and exocytosis
- pinocytosis and phagocytosis

2 **passive transport:**
- osmosis
- diffusion (simple and facilitated).

Links For more information on active transport, see page 51.

> Particles will continue to diffuse from a high concentration to a low concentration until all the particles are evenly spread out.

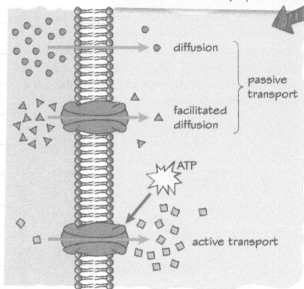

Simple diffusion

Simple diffusion is where there is **passive** movement of substances (not using energy) from a **higher** concentration to a **lower** concentration. A difference between concentrations is known as a **concentration gradient**. For example, during **gas exchange** in the alveoli, oxygen moves from being at a **higher** concentration in the air to a **lower** concentration in the blood.

Osmosis

Osmosis is a special form of diffusion that involves the movement of **solvent molecules** (usually **water**).

There is a net movement of water molecules from a solution with low solute concentration to a solution with a higher solute concentration. If the two solutions are the same concentration, there will be no net movement as they are **isotonic** to each other across a semi-permeable membrane.

Facilitated diffusion

Facilitated diffusion involves molecules which are large or charged. These molecules can only move into or out of a cell with the help of membrane proteins, **down** a concentration gradient. **Channel proteins** open and close to allow specific substances through, and **carrier proteins** aid transport across the membrane.

Now try this

State **one** method of transport that allows oxygen to move in and out of cells.

Active transport

Unlike passive transport, some forms of transport across cell membranes require **energy**. This is called **active transport**, and involves moving molecules **against** a concentration gradient.

ATP

ATP (adenosine triphosphate) is a nucleotide. When ATP is converted to ADP (adenosine diphosphate) it releases **energy** and an inorganic phosphate molecule. This energy is available for cell processes such as muscle contraction.

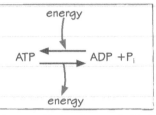

Sodium–potassium pumps

Here, substances move from a **lower** to a **higher** concentration. **Energy** is gained through respiration in the form of **ATP**.

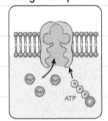 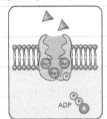 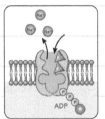 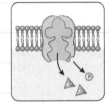

The sodium–potassium pump binds three sodium ions and a molecule of ATP.

The splitting of ATP provides energy to change the shape of the channel. The sodium ions are driven through the channel.

The sodium ions are released on the outside of the membrane and the new shape of the channel allows two potassium ions to bind.

Release of the phosphate allows the channel to revert to its original form, releasing the potassium ions on the inside of the membrane.

Special types of active transport

Large molecules and other particles, such as bacteria, are transported in membrane-bound vesicles. These are active transport processes, requiring energy, supplied by ATP.

Endocytosis (into the cell)

Endocytosis is important because most molecules needed for the cell to survive cannot normally pass through the plasma membrane. **Phagocytosis** and **pinocytosis** are types of endocytosis.

Exocytosis (out of the cell)

In exocytosis, material is exported out of the cell via secretory vesicles. Exocytosis is important in removing waste materials from the cell and in secreting cellular products (e.g. enzymes or hormones).

Phagocytosis/pinocytosis

In phagocytosis, the cell's plasma membrane surrounds a **molecule**, such as a food particle, in the extracellular environment and buds off to form a vacuole that contains the molecule. The molecule is then digested by enzymes.

Pinocytosis is the same as phagocytosis, except the cell's plasma membrane surrounds **droplets of fluid** containing dissolved solutes.

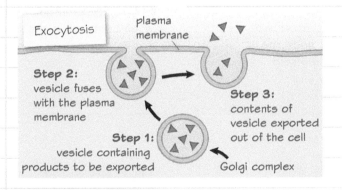 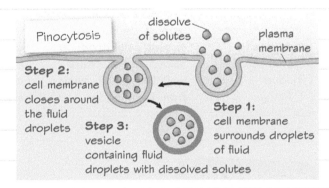

> ## Now try this

Name the molecule used to supply energy in active transport.

Animal tissue types

Tissues are made of specialised cells which work together to perform a particular function.

Cells, tissues and organs

A **cell** is the basic structural and functional unit of an organism.

A **tissue** is a group of cells with similar structures that work together to perform a shared function.

An **organ** is a structure made up of different tissues that work together to perform specific functions.

An **organ system** is a group of organs with related functions.

Tissue types

There are **four** basic types of animal tissue which have **specific** functions within the body:

	Tissue type			
	Epithelial	Connective	Nervous	Muscle
Function	• Lines the body surfaces, cavities and tubules (e.g. kidney tubules). • Important in absorption, secretion and protection.	Supports body parts and connects them together.	• Conducts nerve impulses by reacting to stimuli. • Important in coordinating bodily functions.	• Allows the body to move. • Also allows movement of individual structures within the body.

Examples of specialised animal tissues

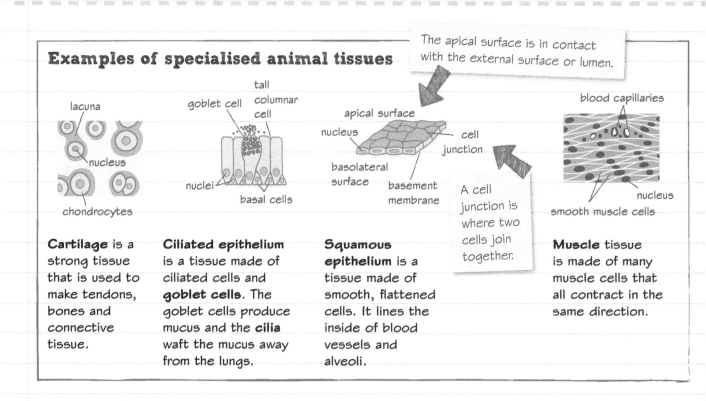

The apical surface is in contact with the external surface or lumen.

A cell junction is where two cells join together.

Cartilage is a strong tissue that is used to make tendons, bones and connective tissue.

Ciliated epithelium is a tissue made of ciliated cells and **goblet cells**. The goblet cells produce mucus and the **cilia** waft the mucus away from the lungs.

Squamous epithelium is a tissue made of smooth, flattened cells. It lines the inside of blood vessels and alveoli.

Muscle tissue is made of many muscle cells that all contract in the same direction.

Now try this

A tissue is a group of cells with similar structures that work together to perform a shared function.

Name the four basic types of animal tissue with specific functions within the body.

Epithelial tissue

Tissues in multicellular organisms are **specialised** to perform a particular **function**. All of the cells in a tissue work together to perform a similar function.

Epithelial tissue

Epithelial tissue is supported by a **basement membrane**, and can consist of a **simple** (single) or **stratified** (multiple) **layer** of cells.

The basement membrane consists of a **network of blood vessels**, providing oxygen and nutrients to the cell and also allowing for absorption. **Secretory epithelia** are specialised to secrete substances such as proteins, for example, glandular cells and goblet cells.

Additional features of epithelial cells

Epithelial cells can be covered in small hair-like projections, called **cilia**. For example, ciliated epithelia in the upper respiratory tract help to move dust and mucus.

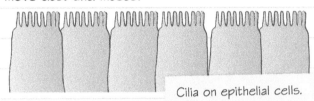

Cilia on epithelial cells.

simple squamous

Location: alveoli
Function: allows substances to diffuse across

simple cuboidal

Location: lines kidney tubules
Function: allows the diffusion and secretion of molecules

Structural differences in epithelial tissue

pseudostratified columnar

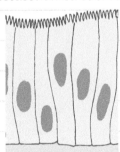

simple columnar

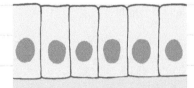

Location: lines the small intestines
Function: allows for nutrient absorption and secretion of substances such as mucus

Location: lines the trachea
Function: allows impurities to be swept towards the throat due to secretion of mucus

Now try this

Tissues in multicellular organisms are specialised to perform a particular function.

What is the name given to epithelial tissue which contains a single layer of flattened cells?
Give an example of where this can be found in the body.

Connective tissue

Tissues in multicellular organisms are specialised to perform a particular function. All of the cells in a tissue work together to perform a similar function.

Connective tissue has large amounts of intracellular matrix and varies considerably in structure and composition. There are three main types of connective tissue: **loose**, **dense** and **specialised**.

Type	Example	Location	Function
Loose connective tissue – loosely packed cells and fibres in a fluid matrix	Adipose	Mainly under the skin around the heart, kidneys and mammary glands	Stores fat and provides insulation
	Areolar	Under all epithelial tissue	Protects organs, blood vessels and nerves; allows passage for nerves and blood vessels through other tissues; gives strength to epithelial tissue
Dense connective tissue – tightly packed cells and fibres with little matrix	Fibrous	Ligaments	Attaches bones to bones and provides support to joints
	Elastic fibres	Tendons	Attaches muscles to bones
Specialised connective tissue	Bone	Skeleton	Forms the skeleton; protects and supports the main organs of the body; anchors the muscle
	Cartilage	Intervertebral discs, between ribs and sternum, external ear	Smooths surfaces at joints and prevents collapse of trachea and bronchi
	Blood	Circulates in the cardiovascular system	Transports substances around the body

Now try this

Tissues in multicellular organisms are specialised to perform a particular function.

Name **two** types of specialised connective tissue.

Muscle tissue

Muscle is an extremely **specialised tissue** containing cells that are capable of contraction. Muscles allow movement of the skeleton. They also have other functions such as causing the heart to beat, allowing the lungs to relax and expand, and moving food through the digestive system.

Myogenesis

Muscle tissue is developed through a process called **myogenesis**. This occurs particularly during embryonic development. Muscle fibres generally form from the fusion of **myoblasts** into multi-nucleated fibres called **myotubes**.

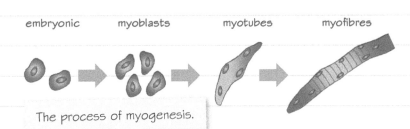

embryonic   myoblasts   myotubes   myofibres

The process of myogenesis.

Myofibrils and sarcomere

- Muscle fibres (muscle cells) are mostly cylindrical **strands of contractile proteins** known as **myofibrils**.

- These **myofibrils** can be broken down into segments called **sarcomeres**, which contain bundles of parallel **actin** and **myosin** filaments.

- The actin and myosin filaments slide in between each other, allowing muscle **contraction and shortening of the sarcomere**.

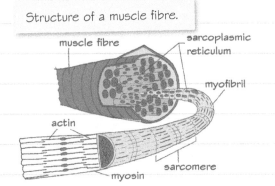

Structure of a muscle fibre.

muscle fibre

sarcoplasmic reticulum

myofibril

actin

myosin

sarcomere

Types of muscle

There are **three** types of muscle. **Voluntary** muscle is used consciously. **Involuntary** muscle is controlled by the **autonomic nervous system**.

Type	Voluntary or involuntary	Location	Appearance	Function
Skeletal	Voluntary	Attached to, support and move the skeleton	**Cylindrical** in shape with **multiple** nuclei arranged around the outside; **striated** in appearance due to the arrangement of the fibres	Movement of the skeleton at joints, they work in antagonistic pairs
Cardiac	Involuntary-myogenic (it can contract without nervous input)	Specialised muscles found in the heart	**Elongated**, branching cells; **one or two** nuclei per cell; **striations** which are characterised by dark and light bands; found with excitatory and conductive muscle fibres, which conduct electrical impulses and control the heartbeat	Contraction of the heart tissue
Smooth	Involuntary	Digestive system, reproductive tracts, the vascular system, bladder and bowels	Cells are **spindle**-shaped with **one** nucleus	Allows hollow organs to contract

Now try this

Muscle tissue is a specialised tissue made up of cells that are capable of contracting.

1 Identify the different muscle tissues that would be in:
   (a) leg   (b) heart   (c) stomach.

2 Give **two** examples of where smooth muscle can be found.

Structure of nervous tissue

Nervous tissue is the main compound of the nervous system and creates **nerve impulses** that travel from one part of a body to another in a few milliseconds. Nervous tissue contains **two** main types of cell, known as **neurons** and **glial cells**.

Neurons

Neurons are specialised cells that conduct information (electrical impulses).

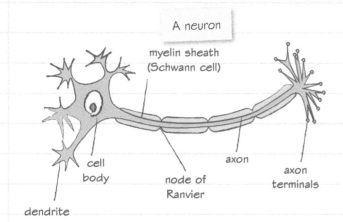

A neuron

myelin sheath (Schwann cell)

cell body

dendrite

node of Ranvier

axon

axon terminals

Links See page 71 for more on how nerve impulses travel.

Glial cells

Glial cells provide **protection** and **support** to nervous tissue. These cells are in direct contact with neurons and often surround them. There are many types of glial cells, such as **astrocytes**, **oligodendrocytes**, **Schwann cells** and **satellite cells**.

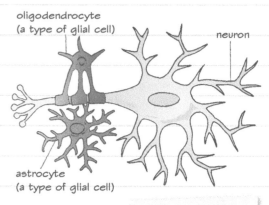

oligodendrocyte (a type of glial cell)

neuron

astrocyte (a type of glial cell)

Types of glial cells.

Types of neuron

There are different types of neuron, which include **sensory** (afferent), **motor** (efferent) and **interneurons** (relay neurons). These three types of neurons form the entire nervous system and have specific relationships with each other. They are found in either the central nervous system **CNS** or the peripheral nervous system **PNS**.

Neuron type	Function	Located
Sensory neurons	Responsible for transmitting impulses from sensory organs to the central nervous system	PNS
Motor neurons	Send impulses from the central nervous system to specific muscles or organs	CNS
Interneurons	Transmit impulses between sensory and motor neurons **within** the brain or spinal cord	CNS

Myelin sheath

A **myelin sheath** surrounds most sensory and motor neurons. It acts as an electrical insulator (like a plastic coating around a wire) which helps speed up the conduction of action potentials and protects the neuron from damage.

Interneurons are mainly **unmyelinated**; this is believed to help them be more sensitive to any changes in their surroundings.

A **synapse** is a junction between nerve cells. **Neurotransmitters** (chemical messengers) transmit the signal across the gap between the nerve cells by diffusion. Their release is triggered by an action potential in the originating neuron.

Now try this

1 Name the **three** types of neuron.

2 Explain why interneurons do not have a myelin sheath.

Fast and slow twitch muscle fibres

There are two types of muscle fibres: **slow twitch (type I)** and **fast twitch (type II)**. All muscles have a mix of both slow twitch and fast twitch fibres. The **combination** of slow and fast determines how good an animal is at physical activities.

Slow twitch fibres

Slow twitch fibres contract slowly but keep going for a long time. They can work for a long time without getting tired.

Animals with a lot of **slow twitch** muscle fibres might be good at **long distance** running as the muscle fibres contract slowly but can keep going longer.

Slow twitch fibres are **red** in colour as they contain **many blood vessels**.

Slow twitch fibres need a lot of **oxygenated** blood to contract the muscles and have a **high** density of **mitochondria** for **aerobic respiration**.

Wolves are an example of an animal with a lot of slow twitch muscle fibre.

Fast twitch fibres

Fast twitch fibres contract quickly but also rapidly **fatigue** as they consume a lot of energy.

Animals with a lot of **fast twitch** muscle fibres might be good **sprinters** or good at **jumping**, as their muscles contract quickly.

Fast twitch muscle fibres are lighter in colour than slow twitch muscle fibres. This is because they have a **lower density of capillaries** as they do not rely on oxygen.

Fast twitch fibres use **anaerobic respiration** to allow a rapid generation of **energy** (adenosine triphosphate – ATP) and have a **low** density of **mitochondria**.

Cheetahs are an example of an animal with a lot of fast twitch muscle fibre.

Structure of muscles.

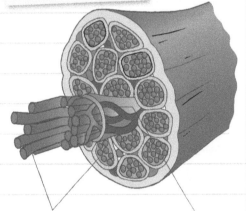

fast twitch and slow twitch muscle fibres are mixed together in a bundle

oxygen is delivered to the muscle fibres via the blood vessels

Aerobic respiration

Aerobic respiration is the release of energy from the breakdown of glucose by **combining with oxygen**. Aerobic respiration is **more efficient** than anaerobic respiration and happens in the **mitochondria** of cells. Energy is used by animals to enable the muscles to contract and allow movement.

Anaerobic respiration

Anaerobic respiration is respiration **without oxygen**. It is a **short-term** energy production method that animals use when they cannot exchange enough oxygen to carry out aerobic respiration (for example, during strenuous exercise). Anaerobic respiration is **less efficient** than aerobic respiration and happens in the cytoplasm of cells. It leaves a poisonous chemical, **lactic acid**, in muscles. This can stop muscles from working well and cause pain.

Links See page 83 for more information on aerobic and anaerobic respiration.

Now try this

Fast twitch muscle fibres provide muscles with the power needed for sprinting over short distances.

1 Give **three** differences between fast twitch and slow twitch muscle fibres.
2 Describe how oxygen is delivered to muscle fibres.

Muscle contraction

Muscles contract through the interaction of two key proteins, **actin** and **myosin**, using **ATP** as an energy source.

Mechanism of muscular contraction

1. For contraction to occur, part of the myosin molecule (head group) attaches to a binding site on the actin filament; a cross-bridge is formed.

2. The head group bends, pulling the actin filament along, and ADP and P_i are released – the power stroke.

3. The attachment of a new ATP molecule to the myosin head breaks the cross-bridge.

4. The head group moves back to its original confirmation as ATP is hydrolysed to ADP and P_i, so another cross-bridge can be formed.

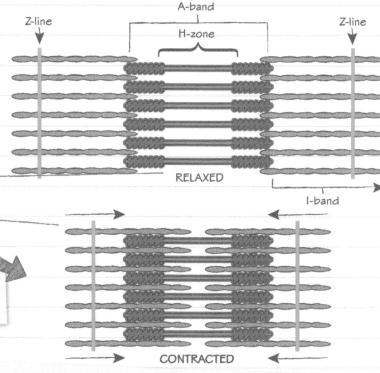

Contraction cycle continues if ATP is available and Ca^{2+} level in the sarcoplasm is high

The binding of Ca^{2+} to troponin causes the myosin binding sites on the actin filament (hidden by tropomyosin) to be revealed so the '**power stroke**' can proceed.

The sliding filament model

Myosin – thick protein filament, each molecule consisting of a tail and two head regions.

Actin – thin filament of globular subunits, twisted like a double strand of beads.

All muscle cells produce a force on contraction because they contain filaments made of the proteins actin and myosin.

Z-line   A-band   Z-line
H-zone
RELAXED
I-band
CONTRACTED

ATP is unstable and only small amounts of it exist in cells at any one time. Contraction of muscles uses lots of ATP and the ATP immediately available is used up after 1–2 seconds.

Antagonistic pairs

The skeletal muscles in animals work in **antagonistic pairs**. For example, when the quadriceps muscles (front of the leg) contract the leg straightens. To bend the leg, the quadriceps muscles relax and the hamstring and adductor muscles on the back of the upper leg contract.

Calcium plays an important role in muscle contraction.

Explain how a lack of Ca^{2+} would affect muscle contraction.

Functions of the skeleton

The skeleton provides a framework for the body in mammals and birds and has several functions.

The skeleton

There are **five** main functions of the skeleton.

1 **Support:** it supports softer tissue and provides points of attachment for most skeletal muscles.

2 **Protection:** it reduces the risk of injury by providing mechanical protection for the body's organs.

3 **Movement:** muscles are attached to bones and when they contract the bones move.

4 **Blood production (haematopoiesis):** red blood cells (erythrocytes), which carry oxygen, and white blood cells (leucocytes), which protect against infection, are produced in the bone marrow of some bones.

5 **Storage of minerals:** bones store minerals, including phosphorus (P) and calcium (Ca), which are released into the blood when required.

Skeleton in a four-legged mammal

The vertebral column is divided into five different sections: cervical, thoracic, lumbar, sacrum and caudal.

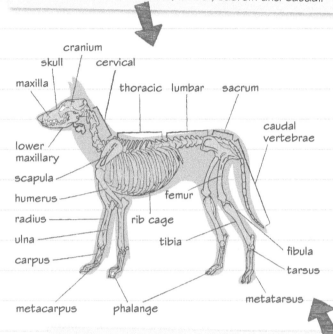

You need to remember all the bones in this diagram. Cover up the labels and see how many you can remember.

Axial and appendicular skeleton

The **axial skeleton** is the bones that make up the head and trunk of the body.

The **appendicular skeleton** consists of the bones of the upper and lower limbs (**limb bones**), and the bony girdles that support them on the body trunk. These bones enable the body to move. They also protect some organs.

The bones of the appendicular skeleton are called **appendage bones**, because they are appendages of the axial skeleton.

☐ Axial
■ Appendicular

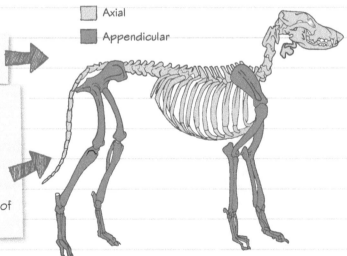

Now try this

The mammalian skeleton consists of both an axial system and an appendicular system.

Which of the following are parts of the axial skeleton?

(a) cranium and vertebrae   (b) humerus and vertebrae   (c) scapula and femur.

Classification, structure and function of bones

Bones are classified by shape, function and location.

Bone classification

There are **five** types of bone.

Type	Shape	Example		Function	Location
Long	Longer than wide	Femur in a dog		Act as levers; aid locomotion and support	Bones of the limbs (e.g. femur, humerus, tibia)
Short	Equal dimensions	Carpal in a cat paw		Absorb impact	Bones of the feet or paws
Flat	Strong, flat plates of bone	Scapula in a horse		Protect organs; muscles attach to them	Bones of the pelvis, cranium (skull) and scapula (shoulder blade)
Sesamoid	Usually short irregular bones, embedded into a tendon	Patella (knee cap) in a dog		Reduce friction	Bones of the fetlock in a horse
Irregular	Odd-shaped bones	Vertebra in a giraffe		Protection, support, anchor points	Bones of the vertebral column

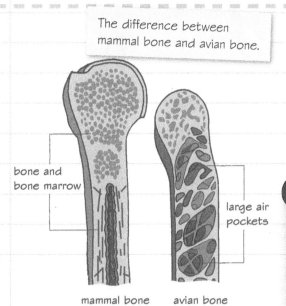

The difference between mammal bone and avian bone.

bone and bone marrow

large air pockets

mammal bone   avian bone

Adaptation of mammalian and avian bones

The differences between mammalian and avian bones are mainly related to birds having wings and, in most cases, being able to fly.

The bones are pneumatised: they have **air pockets** and are reinforced with a **honeycomb** structure which makes the bones lighter but strong. Also, the collarbone of the bird is fused for stability.

Now try this

An animal's skeleton consists of different types of bones. The different bones all have different functions.

Give an example of a bone of each of the following types:

1 short   **2** sesamoid   **3** irregular.

Joints and muscles

Joints and muscles work together to move the body.

Joints

A joint is the point where two or more bones meet. Some joints are freely movable, some joints only move slightly while others are fixed and so do not move at all.

Fibrous joints are fixed by fibrous connective tissue that allows no movement. These are mainly found in the skull.

Cartilaginous joints are held together by cartilage and only allow slight movement. These can be found in the spine and ribs. Cartilage is a good shock absorber.

Synovial joints are freely moveable and occur where two bones meet.

Synovial joint components

Hyaline cartilage: reduces friction and acts as a shock absorber.

Ligament: joins bone to bone and stabilises the joint.

Tendon: joins muscles to the bones and enables movement.

Synovial membrane: produces synovial fluid.

Synovial fluid: lubricates the joint.

Fibrous joint capsule: an envelope around the synovial joint.

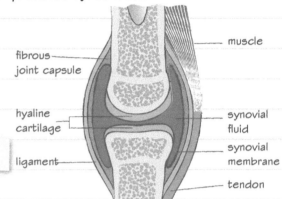

A synovial joint.

Types of synovial joint

Joint type	Movement at joint	Example
Hinge	Flexion/extension	Elbow/knee
Pivot	Rotation of one bone around another	Top of the neck (atlas and axis bones)
Ball and socket	• Flexion/extension • Adduction/abduction • Internal and external rotation	Shoulder/hip
Gliding	Gliding movements	Intercarpal joints
Condyloid	• Flexion and extension • Abduction and adduction	Wrist, phalanges
Saddle	Most movements apart from rotation	Thumb

Muscles

Muscles are used for:

- **locomotion** – moving limbs
- **organ movements** – for example, contraction of the intestines and pumping of the heart
- **posture** – holding body position
- **heat generation** – shivering.

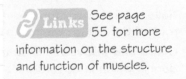 See page 55 for more information on the structure and function of muscles.

See page 55 for more information on the structure and function of muscles.

Now try this

Give the names of the **six** types of synovial joint.

The integumentary system

The **integumentary system** protects the body from damage, such as loss of water or abrasion from outside. The system is made up of the skin and its appendages (including hair, scales, feathers, hooves, and nails).

Skin structure

Skin covers the external surface of the body. It is made up of three layers:

(1) Epidermis – waterproof, outer layer which keeps pathogens out of the body. It is elastic and gives skin its colour.

(2) Dermis – middle layer which contains hair follicles, nerves, blood vessels and glands to help regulate body temperature.

(3) Subcutaneous layer – made of connective tissue and has a high amount of fat which provides insulation and shape.

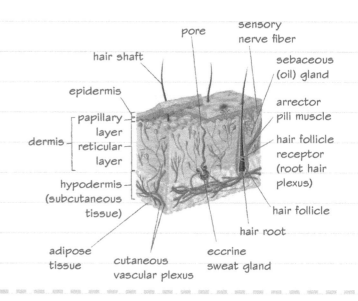

Skin glands

Skin contains different **exocrine glands**, each secreting different substances:

- **sudoriferous** – secrete sweat
- **sebaceous** – secrete sebum which keeps hair and skin lubricated
- **ceruminous** – specialised sudoriferous glands which secrete ear wax.

Hair

Hair is made of **keratin**. In animals, guard hair and downy hairs grow from the same **follicle**.

Downy hair – dense, fine, soft hairs which lie close to the skin making up the undercoat. Good for insulation.

Guard hair – coarser, thicker, longer hairs make up the top coat. Help prevent injuries to the skin and influences the animal's appearance.

Each guard hair is surrounded by 6–12 downy hairs.

Nails, claws and hooves

Nails – a nail matrix forms on top of the dermis. The root of the nail grows from a thick layer of matrix.

Claws and hooves – made of a section of the dermis which is covered in a layer of keratinised epithelium.

Deer with young antlers covered in velvety skin.

Horns and antlers

Horns and **antlers** are made from different things in different species.

Hollow horns (e.g. goat, sheep) are made from a cone of keratin which surrounds a mass of bone. Rhino horns are different because they are made from keratinised cells without a core of bone.

Antlers are bone. Young antlers are covered in velvety skin which is rubbed off as the antlers develop. The bone turns into compact bone tissue as they mature.

Now try this

Arctic foxes have adapted to survive in cold climates.

Explain the role of the skin and its appendages in maintaining the fox's internal body temperature.

Structure and function of feather types

Feathers are made of a particular type of **keratin** called **beta keratin**. You need to know about the different types of feather.

Feather types

Contour – give the bird its shape and colour; they can also help insulate. 	**Down** – soft and fluffy, these help insulate the bird. 	**Semiplume** – found underneath contour feathers, used for insulation. 
Flight – found in the wings and tail; these give strength for flight. 	**Bristle** – found around the eyes and mouth of insect-eating birds; they protect the eyes and help funnel food into the mouth.	**Filoplume** – very small and attached to nerve endings; they send information to the brain about feather alignment. 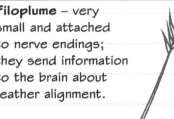

Feather functions

1. **Flight** – flight feathers help the bird stay airborne, contour feathers make the bird streamlined.

2. **Protection** – feathers protect underlying skin and areas around the eyes and mouth.

3. **Warmth** – feathers are insulating; the numbers of feathers a bird has can change throughout the year, e.g. redpolls have 30% more feathers in winter.

4. **Stealth** – in some species, for example, owls, flight feathers are adapted for silent flight.

5. **Displaying** – as part of mating behaviours.

6. **Avoiding predation** – some birds, for example, the dark-eyed junco, have brightly coloured feathers which they can flash to distract predators. Others use their feathers as camouflage.

7. **Walking** – ptarmigans have feathers on their feet; this helps them walk on snow.

Feather structure

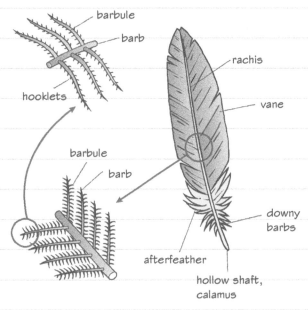

Now try this

Recent studies on emperor penguins have discovered that a large amount of tiny filoplumes have been found, not for the purpose of flight, but to create a waterproof covering and additional insulation.

Apart from filoplume feathers, which other type of feathers do these penguins have?

Methods of locomotion

The skeleton gives rise to different types of **locomotion (movement)** exhibited by different species. This aids their **survival**.

A chimpanzee using its long limbs and strong grip to climb trees.

Climbing
Adaptations: long limbs, opposable hands and feet; often a prehensile tail; keratinised nails and claws to improve grip, e.g. monkeys and apes

Hopping/leaping
Adaptations: strong muscles, elastic tendons, and long and powerful rear limbs, e.g. kangaroos

Gliding
Adaptations: lightweight skeleton with a **patagium** (a fold of skin between the forelimbs and hindlimbs); some have a tail or cartilaginous wrists to direct movement, e.g. flying squirrels

A dolphin using its tail to propel itself out of the water.

Types of movement

Swimming
Adaptations: caudal vertebrae to give a long tail for propulsion, fins, e.g. dolphins

Running
Adaptations: long limbs, lightweight skeletons; some have flexible spine or absence of some bones to allow for more movement (collar bone in cheetah, distal limb bone in horses); some have tiny hairs on the pads of the feet to help grip on smooth surfaces, e.g. polar bear

The integumentary system shows differences in the type and extent of keratinisation depending on the type of locomotion the animal uses.

Powered flight
Adaptations: long bones to form wings, lightweight bones with a honeycomb structure, beaks lighter than mandible; feathers for flight or to provide waterproofing for diving, e.g. birds

Remember you will need to be aware of different skeletal adaptations various species of mammals and birds have, and how these help them to move. See page 65 for examples.

🔗 **Links** See page 62 for more on the integumentary system.

Now try this

Cheetahs have several musculoskeletal adaptations for running.

Describe **two** of these adaptations.

Musculoskeletal adaptations and disorders

Every species has a different body shape and physical ability depending on what they require to survive, and this has **adapted** through evolution.

Adaptations of mammals

Long neck to reach food and monitor the area for predators

Ball and socket joints between cervical vertebrae and first and second thoracic vertebrae for greater **flexibility**

Dense limb bones with thick walls to support weight

Caudal vertebrae form the tail which helps with **balance**

Skeleton of a giraffe

Forelegs ~10% longer than hind legs to support the neck, along with much muscle development in the neck and shoulders

Hoof to support weight, protect the bone and offer more traction when running

Ligaments not muscles in the lower limbs to increase **stability** and energy efficiency

Adaptations of birds

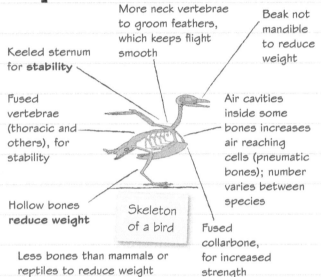

More neck vertebrae to groom feathers, which keeps flight smooth

Beak not mandible to reduce weight

Keeled sternum for **stability**

Fused vertebrae (thoracic and others), for stability

Air cavities inside some bones increases air reaching cells (pneumatic bones); number varies between species

Hollow bones **reduce weight**

Skeleton of a bird

Fused collarbone, for increased strength

Less bones than mammals or reptiles to reduce weight and increase stability

Musculoskeletal disorders

Musculoskeletal disorders can be caused by problems with the bones or joints. They can affect the body's ability to move and function normally. Some species and breeds are more susceptible than others; in domesticated species this is mostly due to **inbreeding**.

In dogs, **hip dysplasia** is caused by malformation of the hip joint, which deteriorates as it is not able to work properly. Symptoms include loose joints, reduced activity, difficulty running or jumping, swaying gait, pain and decreased movement in the joint.

Osteoarthritis is very common and can give rise to other diseases. It affects the cartilage of joints, leading to bone rubbing on bone. This can alter the shape of the joint. There are many causes, such as injury, incorrect weight and insufficient nutrition. Symptoms include stiffness, inability to rise and uneven gait. Osteoarthritis is progressive, leading to reduced mobility which can then lead to obesity and increased stress on the joints.

Osteochondritis occurs when the normal development of cartilage to bone is stopped. The cartilage is then more susceptible than bone to mechanical damage. This can be caused by genetics or trauma. Symptoms include: lameness, which becomes worse after exercise; inability to weight-bear on affected limbs; pain; muscle wastage. A long-term management plan will be needed to manage the illness. This may include restricting exercise, although quality of life may always be reduced.

Now try this

A vet completes a health check on a senior Springer spaniel and notices stiffness and limping and an onset of arthritis.

1   State the cause of osteoarthritis.

2   Describe the consequences of osteoarthritis with regard to the animal's physical condition.

Nutrients and digestion

Animals are **heterotrophic**. They obtain the nutrients they need to survive from plants, or other animal tissue. These nutrients are then processed by the digestive system.

Nutrients

There are **seven groups** of nutrients; each group has a different function.

Nutrient	Functions in the body
Proteins	Proteins are necessary for the growth and maintenance of all body cells and tissues. Enzymes are proteins, as are some hormones, and proteins also form some transport molecules, such as haemoglobin. They are also a source of energy.
Carbohydrates	Carbohydrates supply energy to the cells.
Lipids (fats)	Fats supply and store energy, and insulate, support and cushion organs. They are also involved in the absorption of fat-soluble vitamins.
Vitamins	Vitamins promote specific chemical reactions within cells.
Minerals	Minerals are used for the growth and maintenance of bones and teeth. They are involved in osmoregulation, nerve transmission, muscle contraction, transport systems and other functions.
Fibre	Adequate dietary fibre is needed to increase bulk and water in the intestinal contents to promote and regulate normal bowel function and transit times.
Water	Water makes up 50–70% of body weight, and provides a medium for chemical reactions. It also transports chemicals, regulates temperature and removes waste products.

Digestion

In order for food to be digested there are **two processes** that occur:

- **Mechanical digestion** – where large pieces of food are ingested and physically broken down into smaller pieces.

- **Chemical digestion** – where larger molecules of food are broken down into smaller molecules by enzymes, acid and bile. For example, **amylase** is an enzyme found in saliva and the stomach; it is responsible for breaking down starch into short-chain carbohydrates.

Monogastric digestion

Food enters the mouth and is **mechanically** broken down by the teeth and tongue. Saliva is added, which starts the **chemical digestion** of food. The food is swallowed and moves down the **oesophagus** to the **stomach**. In the stomach, **hydrochloric acid** and **enzymes** (proteases and lipases) are added, which break down the food even further to form **chyme**. When it enters the **duodenum**, **bile** (made by the liver and stored in the gall bladder) **emulsifies fats** and **neutralises stomach acid,** and more enzymes (made by the **pancreas**) are added.

In the **jejunum** and **ileum**, nutrients are absorbed and taken to cells for use or storage. In the **large intestine**, water and some water-soluble vitamins are absorbed through the villi. The villi provide a larger surface area for absorption. They contain a network of capillaries to allow for the effective diffusion of molecules. Waste material is stored in the **rectum** until it leaves the body via the **anus**.

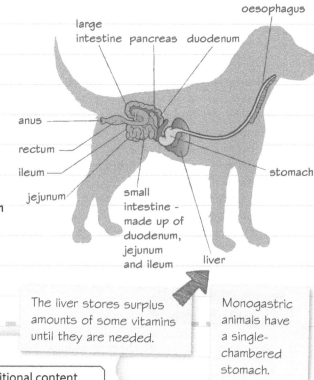

The liver stores surplus amounts of some vitamins until they are needed.

Monogastric animals have a single-chambered stomach.

Now try this

Researchers are analysing animal feeds to determine their nutritional content.

Give **four** nutrient groups that are essential in an animal's diet.

Specialised digestive systems

Digestion is a complex process which differs depending on the animal's diet. Digestive systems are **adapted** to help animals get nutrients from their food.

Hindgut fermenters

Some **herbivores** eat mostly high fibre plant material. **Fibre** is difficult to digest, so their digestive system needs to be **adapted** to their diet.

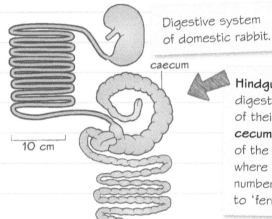

Digestive system of domestic rabbit.

caecum

10 cm

Hindgut fermenters digest the majority of their food in their **cecum** and at the start of the large intestine, where there are large numbers of **microbes** to 'ferment' the food.

Bird digestion

Birds do not chew. Instead they use their tongue to move food to the back of their mouth where it is swallowed. Food moves from the oesophagus to the **crop** (a temporary storage area), before passing to the **proventriculus** where digestive enzymes and hydrochloric acid are added. Then food moves to the **ventriculus** (the 'gizzard') where it is mechanically digested. Where the small and large intestines join, there are two **ceca** (pouches) where some water is reabsorbed and the food is fermented. The **cloaca**, at the end of the digestive system, mixes digestive and urinary waste and expels them as one substance.

Ruminant digestion

Ruminants are specialised **foregut fermenters**. Their stomach consists of four parts: **rumen, omasum, abomasum** and **reticulum**.

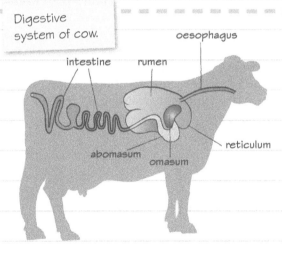

Digestive system of cow.

oesophagus

intestine   rumen

abomasum   omasum   reticulum

1 Fermentation occurs in the **rumen** in an anaerobic environment where microbes secrete enzymes such as cellulase to breakdown cellulose. Saliva helps to ensure there is a suitable environment for micro-flora (bacteria) to survive. The reticulum aids the rumen and helps form the undigested feed into **cud**. **Retro** (or reverse) **peristalsis** allows for feed to be brought back up to allow for it to be mechanically broken down by chewing and then swallowed, allowing further digestion.

2 The **reticulum** filters the food in the rumen, allowing only the parts which are sufficiently broken down to move on to the omasum.

3 In the **omasum,** some water and salts are absorbed, then the remaining matter is passed to the abomasum.

4 In the **abomasum**, the true stomach, digestive enzymes and acid are added.

Now try this

1 State the food stuff ruminants can digest that non-ruminants cannot.

2 Complete the table below for the parts of the stomach of a ruminant.

Part of stomach	Function
Omasum	
	Catches foreign objects and works alongside the rumen
Rumen	
	Where enzymes break up food – the 'true stomach'

Oral cavity adaptations

Some species have special adaptations for digestion. Differences include **adaptations of dental formulae** and **stomach formation**. Species vary in the adaptations for efficient digestion of their teeth and digestive system.

Herbivore dentition

Herbivores include **ruminants** such as cows, sheep and giraffes, and have a diet comprised of **plant** material.

Instead of upper incisors, they have a **dental pad**. They grasp plants with their tongue and pinch it off with the lower incisors against the dental pad. They do **not** have canine teeth. They thoroughly **grind** food and mix it with large amounts of saliva. Their molars continually grow as they become worn by the silica in the plant material they eat.

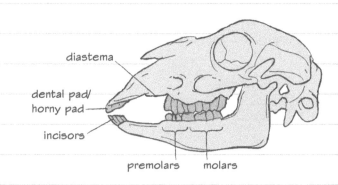

Carnivore dentition

Carnivores, such as dogs, tigers and lions, have a diet that consists of **meat**. They have teeth specialised for different tasks. To enable them to rip meat, they have teeth known as **canines**. Paired modified molars known as **carnassial teeth** allow the shearing of meat. This is more efficient than tearing. They do not have the ability to digest cellulose as they lack the enzyme cellulase.

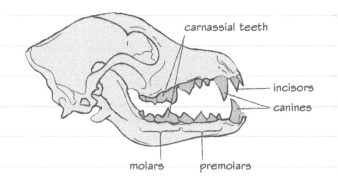

Oral cavity of birds

Bird's beaks are made of bone, keratin and blood vessels. They grow continually throughout a bird's life and are worn down by grooming, feeding, climbing and rubbing. The shape of the beak is adapted to the type of food the bird eats, for example, a pelican has a large lower beak for scooping fish. There are no teeth inside a bird's mouth as they do not chew their food.

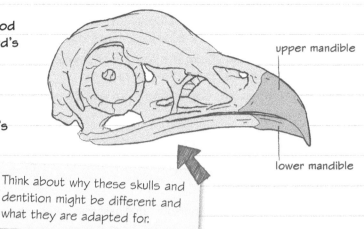

Think about why these skulls and dentition might be different and what they are adapted for.

Remember the diet of a wolf and how this has influenced its dentition.

Now try this

Explain a feature of wolf dentations which help them eat their food.

Regulation of blood glucose

Glucose is needed by cells for respiration. It is important that the concentration of glucose in the blood is maintained at a constant level, as too much or too little can lead to coma and even death.

How blood glucose is regulated

Blood glucose levels are regulated by the pancreas in a **homeostatic** process. The **pancreas** detects blood glucose concentrations, then releases **hormones** which alter the amount of **glucose** in the blood.

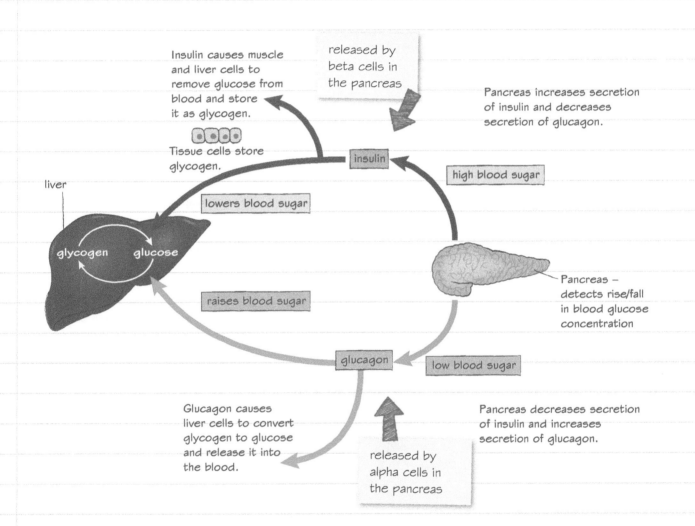

Insulin causes muscle and liver cells to remove glucose from blood and store it as glycogen.

Tissue cells store glycogen.

released by beta cells in the pancreas

Pancreas increases secretion of insulin and decreases secretion of glucagon.

insulin

high blood sugar

lowers blood sugar

liver

glycogen   glucose

raises blood sugar

Pancreas – detects rise/fall in blood glucose concentration

glucagon

low blood sugar

Glucagon causes liver cells to convert glycogen to glucose and release it into the blood.

released by alpha cells in the pancreas

Pancreas decreases secretion of insulin and increases secretion of glucagon.

A dog has recently been diagnosed with diabetes, caused by a problem in the mechanism to reduce blood glucose.

Explain the normal mechanism for detecting increased blood glucose levels and reducing them.

Digestive disorders

Digestive disorders can alter the digestion or absorption of food, or affect the passage of food through the digestive tract. These can affect an animal's stomach and intestines, resulting in pain and other problems.

Healthy digestion

It is essential that every animal has a healthy digestive system to be able to use the nutrients from their food to build and repair tissues and obtain energy. **Digestive disorders** can lead to dehydration, acid–base and electrolyte imbalances, and malnutrition. Some of the most common digestive disorders include ingestion of foreign bodies, bloat, and sickness and diarrhoea.

Ingestion of foreign bodies

Foreign bodies can cause obstructions in the digestive system. These can **slow** the digestive process, lead to regurgitation of food or **block** the digestive system completely.

Some **foreign bodies**, such as nails or tools, can puncture the digestive system, leading to infection and death.

Treatment may involve surgery to remove the item.

Ruminant bloat

Bloat can occur in any ruminant, but is most common in **cattle**. It is caused by a build-up of gas in the rumen and can be fatal. Frothy bloat is caused by foam in the rumen trapping the gas; this is more common in animals fed a high protein, low roughage diet. Free-gas bloat is caused by obstructions in the oesophagus.

Surgical procedures are needed to relieve the bloat.

Sickness and diarrhoea

These can be caused by a number of things such as:

- ✓ unsuitable amounts of food
- ✓ contaminated food
- ✓ unsuitable food types
- ✓ stress
- ✓ pathogens.

Treatments will vary depending on the cause.

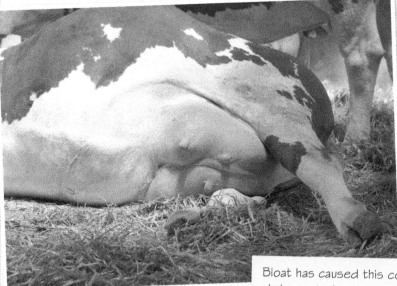

Bloat has caused this cow's abdomen to become swollen.

Now try this

Hardware disease is seen in cattle that have ingested tools. Explain the impact this may have on the animal's digestive system.

Action potentials

Neurons send nerve impulses, known as **action potentials**. Action potentials involve changes to the concentration of **ions** such as **potassium (K^+)** and **sodium (Na^+)** on both the inside and outside of the axon.

Action potential

When information gets sent through a neuron, we call this an 'action potential'. During an action potential, there is a **movement of ions across the membrane**, initiating a **nerve impulse**.

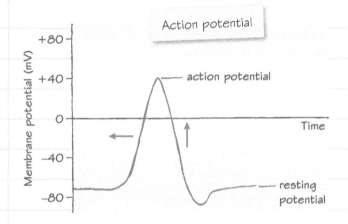

Stages of an action potential

A **stimulus** causes voltage-gated Na^+ channels to open and Na^+ ions diffuse in. This begins to increase the **positive charge** inside the cell.

- If the **threshold** is reached, an **action potential** is **triggered**. This **opens** more Na^+ **channels** and *depolarisation* occurs. This is where the cell becomes **more positively** charged on the **inside** than out.

- **At the peak voltage** of around **+40 mV**, Na^+ channels **close** and voltage-gated K^+ channels **open**. K^+ ions diffuse out, causing *repolarisation* of the cell.

- This causes *hyperpolarisation* of the neuron. More K^+ ions are on the outside than Na^+ ions are inside.

- A **refractory period** allows the sodium-potassium pump to return K^+ ions to the **inside** and Na^+ ions to the **outside**, returning the neuron to its **normal polarised** state.

Impulses travel along an axon insulated with **myelin**. The insulation wraps around the axon with gaps in between called **nodes**. An action potential 'jumps' across these gaps in **salutatory conduction**.

Resting potential

At rest, inside a neuron, there is a negative charge compared to outside. This is caused by the distribution of Na^+ and K^+ ions which is regulated by the **sodium–potassium pump**.

The difference between the charges inside and outside the neuron gives the membrane a potential **voltage** of around **−70 mV**.

At the end of a neuron

At the end of an action potential, the impulse will be transmitted to the next set of cells. The way this happens depends on the type of cells:

Other neurons – the junction between two neurons is called a synapse

Muscle tissue – a neuromuscular junction

Glands – a neuroglandular junction.

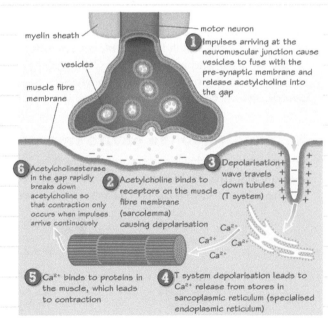

Transmission across the neuromuscular junction involves **calcium ions (Ca^{2+})**. After the muscle contraction is over, the calcium channels close and calcium is removed from the muscle cells.

Now try this

When a stimulus reaches a resting neuron, the neuron transmits the signal as an impulse and this is called an action potential.

Describe the four stages of the action potential.

The nervous system

The nervous system is very important as it allows our bodies to respond to the environment.

The central nervous system (CNS)

The **brain** and **spinal cord** form the **central nervous system** (CNS). The **CNS** controls most of the functions and responses in the body.

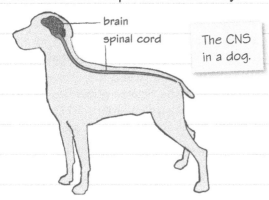

The CNS in a dog.

The **spinal cord** is a long bundle of nerves which branches out between the **vertebrae** to different parts of the body. This connects the brain to the **peripheral** nervous system.

The CNS is surrounded by a shock absorbing fluid called **cerebrospinal fluid** (CSF). The CNS is kept separate to the rest of the body fluids by the blood–brain barrier. This is made of specialised cells which line the blood vessels of capillaries in the region of the CNS, giving some protection from **changes to blood composition**, **toxins** and **pathogens**.

The peripheral nervous system (PNS)

The **nerves** that connect the **spinal cord** to the rest of the **body** are known as the **peripheral nervous system** (PNS). The **PNS** is made up of receptors, sensory (afferent) and motor (efferent) neurons. Parts of the PNS form the **somatic nervous system**; this detects and responds to **stimuli** outside the body.

When a **receptor** is stimulated, a **signal** is sent to the **CNS**.

Spinal nerves are connected to the spinal cords and **cranial** nerves to the brain.

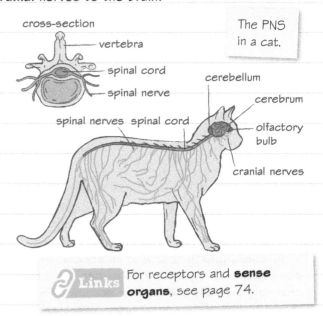

The PNS in a cat.

Links For receptors and **sense organs**, see page 74.

Responding to the environment

Responding to stimuli can be **voluntary** (the brain is involved in determining the response) or **involuntary** (the brain is not involved in determining the response). Sometimes it could take too long for the information to go to the brain to be processed.

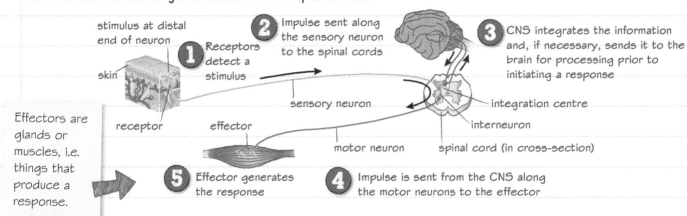

1 Receptors detect a stimulus — stimulus at distal end of neuron — skin — receptor

2 Impulse sent along the sensory neuron to the spinal cords — sensory neuron

3 CNS integrates the information and, if necessary, sends it to the brain for processing prior to initiating a response — integration centre — interneuron — spinal cord (in cross-section)

4 Impulse is sent from the CNS along the motor neurons to the effector — motor neuron

5 Effector generates the response — effector

Effectors are glands or muscles, i.e. things that produce a response.

Now try this

Whiskers are attached to free nerve endings. Explain how signals from whiskers reach the brain.

The autonomic nervous system

The **autonomic nervous system** is part of the **peripheral nervous system** within the nervous system. It is made up of the **parasympathetic** and **sympathetic** divisions. This system is regulated by the **hypothalamus**.

The autonomic nervous system

The autonomic nervous system has two divisions that work together 24 hours a day:

- **Parasympathetic** ('*rest and digest*') restores normal functioning after extra stimulation of the sympathetic system is no longer being carried out.
  The neurotransmitter is **acetylcholine (ACh)**.
- **Sympathetic** ('*fight or flight*') prepares the body for physical activity.
  The neurotransmitter is **noradrenaline (NA)**.

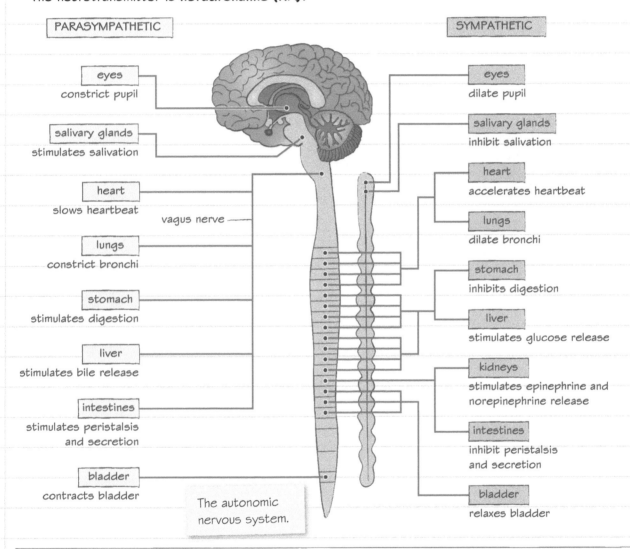

PARASYMPATHETIC

eyes
constrict pupil

salivary glands
stimulates salivation

heart
slows heartbeat

vagus nerve

lungs
constrict bronchi

stomach
stimulates digestion

liver
stimulates bile release

intestines
stimulates peristalsis
and secretion

bladder
contracts bladder

SYMPATHETIC

eyes
dilate pupil

salivary glands
inhibit salivation

heart
accelerates heartbeat

lungs
dilate bronchi

stomach
inhibits digestion

liver
stimulates glucose release

kidneys
stimulates epinephrine and
norepinephrine release

intestines
inhibit peristalsis
and secretion

bladder
relaxes bladder

The autonomic
nervous system.

Different spellings

You may find that spelling of certain hormones varies in different text books or websites. For example: noradrenaline/noradrenalin/norepinephrin/norepinephrine. This is due to different languages spelling it slightly differently.

Now try this

1 What is the neurotransmitter used in the synapses of the parasympathetic nervous system?
2 Which division of the nervous system would be active in a mouse trying to escape a predator?

Receptors and sense organs

The nervous system consists of the brain, spinal cord and sensory organs. The **sensory organs** play a vital role within the system, sensing changes to the environment surrounding the animal.

The sensory organs

Sensory organs contain **receptors** which detect stimuli. Each sense organ detects a particular type of stimuli which depends on the types of receptors present.

Types of receptor:

- **Chemoreceptors** – detect the presence of chemicals
- **Thermoreceptors** – detect changes in temperature
- **Mechanoreceptors** – detect mechanical changes in the environment
- **Photoreceptors** – detect light.

Receptors

Receptors are the parts of the body which **detect stimuli** and change as a result of the stimuli. Receptors are connected to **sensory neurons**. Changes in the receptors can lead to action potentials in the sensory neurons which are attached.

A bush viper with eyes open.

Auditory – hair cells in the inner ear vibrate according to the pitch and volume of sounds. They send impulses via the auditory nerve to the brain for processing.

Somatosensory – the skin contains a variety of receptors which respond to temperature, touch or pressure. Whiskers also contribute to somatosensation.

Visual – photoreceptors in the retinal layer of the eye respond to light.

Links For more information on photoreceptors see page 76.

Sense receptors

A cat's whiskers.

Gustatory – the tongue of many animals contains papillae which contain taste buds. These respond to chemicals in food and drink to give different sensations of taste. Different animals have different taste receptors, so not all animals perceive taste the same way.

Olfactory – receptors in the nasal cavity respond to odour molecules in the air. Some animals, e.g. snakes, have a Jacobson's organ which specifically detects moisture-borne odour molecules. The flehmen response helps animals detect the moisture-borne molecules using their Jacobson's organ.

A chameleon catching a cricket with its tongue.

Some species have other sensory adaptations which help them detect changes in the environment. Some of these adaptations are:
- Lateral line system
- Electroreceptors
- Vibrissae
- Barbels.

Now try this

State the main somatosensory organ in animals.

The structure of the eye

The eye is a complex organ and is vital in many species' survival. The eye's **receptors** are sensitive to light stimuli and send visual signals to the brain through the nervous system.

Parts of the eye

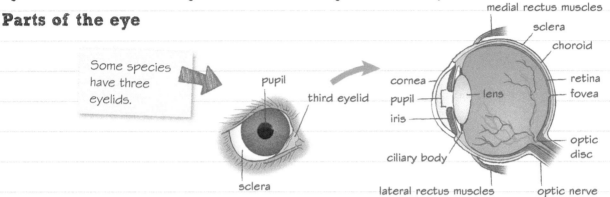

Some species have three eyelids.

pupil
third eyelid
sclera

medial rectus muscles
sclera
choroid
cornea
pupil
iris
lens
retina
fovea
ciliary body
optic disc
lateral rectus muscles
optic nerve

Name	Description	Purpose
Eyelid	Fold of skin which covers the eye	Cleaning and protecting the eye
Cornea	Transparent covering on the eye	Protecting the eye; helping converge (bend and join) light rays that enter the eye
Sclera	White part of the eye made of tough collagen fibres	Providing protection and support
Retina	Layer of light-sensitive cells (photoreceptors) at the back of the eye	Detecting light rays that are focused by the cornea and lens
Choroid	Layer of blood vessels with a black pigment	Nourishing the eye and absorbing light
Iris	Ring of muscles around the pupil, coloured which provides eye colour	Controlling the amount of light entering the eye
Pupil	Hole in the middle of the iris	Allowing light to enter the eye
Lens	Transparent, flexible and can change shape due to the **ciliary muscles**	Focusing the light onto the retina and focusing on objects
Ciliary body	Attached to the underside of the lens	Producing **aqueous humour** and helping with focusing by altering the shape of the lens
Vitreous humour	Located behind the lens	Giving the eyeball its shape
Aqueous humour	Found behind the cornea	Helping to keep its rounded shape; providing nutrients to the cornea and lens
Lateral rectus muscles	Found on the lateral side of the eye	Helping orientate the pupil away from the centre of the body
Medial rectus muscle	Found on the medial side of the eye	Helping orientate the pupil towards the centre of the body
Fovea	Spot located in the macula, which has a high density of cone cells	Giving sharp central vision; light is focused onto this spot by the lens
Optic disc	Blind spot where there are no photoreceptors	Where the optic nerves leave the eye
Optic nerve	Bundle of fibres	Relaying information from the retina and fovea to the brain

Now try this

Explain why the aqueous humour is needed to provide nutrients to the cornea.

How the eye works

The structure of the eye enables images to be detected and transmitted to the brain through the **optic nerve**.

The vision process

Light rays are refracted and focused on the **pupil** as they enter through the **cornea**. Muscles in the **iris** contract or relax to control the amount of light passing through the pupil.

→

Behind the pupil is the **lens**, which refracts the light more so it is focused on the **retina**.

→

The photoreceptors in the **retina** respond to the energy from the light and generate **action potentials** which are sent along the **optic nerve** to the brain for processing.

Photoreceptors

These are the **primary layer** to the retina and consist of **rods** and **cones**.

- **Rods:** very light sensitive and can function in the dark. They have a single pigment known as **rhodopsin**, which cannot detect colour and so vision is in shades of grey. They have a **low** level of **visual sharpness** as many rod cells **share** a **connection** to the optic nerve.

- **Cones:** function at a much higher intensity than rods, and allow some species to see in colour as they contain photo pigments (**iodopsin**) which respond to colour. Each cone cell responds to one colour. The range of colours detected depends on the types of cone cells present. Cones give a **high** level of **visual sharpness**, as each cone has a **single connection** to the optic nerve.

Different animals have different types of photoreceptors and in different densities, for example, birds have cones which respond to violet, blue, green and red light, while dogs have cone cells which respond to violet and yellow-green light.

Structure of the retina

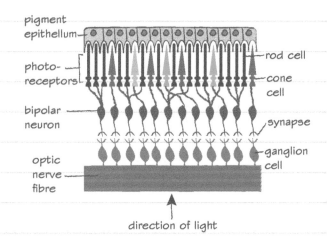

pigment epithellum
photo-receptors
bipolar neuron
optic nerve fibre

rod cell
cone cell
synapse
ganglion cell

direction of light

Use of light

When **light** is detected by **rods, rhodopsin** is broken down into two proteins known as **opsin** and **retinal** via a process known as **bleaching**. This causes **depolarisation**, giving rise to an **action potential** which is transmitted via the **optic nerve** to the brain. Rhodopsin is regenerated slowly. Cone cells provide the visual information when light levels are high, as rhodopsin is unable to effectively regenerate.

UV sensitivity in bird eyes is thought to help them forage, as many fruits, seeds and insects are visible in UV light.

Now try this

1   Where is light focused as it enters the eye?
2   Rhodopsin is broken down into two proteins. Name both.

Eye adaptations

Different species require different levels of vision to enable them to survive in their environment.

Nocturnal animals

Nocturnal animals have a variety of eye adaptations to enable them to see in the dark (not all nocturnal animals have all of these!):

1 Much larger eyes which have **wider pupils**. This allows them to capture more light.

2 A much **higher concentration** of **rod cells** in the retina.

3 **Slit pupils** as they are easier to contract than round.

4 A special layer to their eyes, known as the **tapetum lucidum**. This layer contains a **reflective pigment** which aids sight in dark conditions. It improves vision in low light as it maximises the available light by reflecting it back through the photoreceptors, stimulating more receptor cells.

Pupil shapes

There are many different pupil shapes.

The **pupillary light reflex** adjusts the size of the pupil in response to the amount of light present. This helps protect the photoreceptors from damage.

Predator and prey adaptations

Predator (carnivore)	Prey (herbivore)
Binocular vision – this is where each eye overlaps at the front to create a single 3D image.	**Monocular vision** – each eye works independently and only overlaps over a small area.
Eyes positioned on **front of face**, narrower field of vision.	Eyes positioned on **side of head**, wider field of vision – good for detecting predators.
Able to focus as have **better depth perception**.	**Poor depth perception**.

For example, owls have a visual field of around 110°. Their eyes are elongated so they cannot rotate in the socket; owls need to turn their head instead. They have large numbers of rod cells which, along with a wide range in pupil size, helps them hunt in low light levels.

For example, sheep have a visual field of over 270°. Their eyes are also high up in the skull which gives them better vision when grazing, while the rectangular pupils increase the accuracy of depth perception while grazing.

Now try this

1 What type of vision do lions have?
2 State **two** advantages and **two** disadvantages for this vision type.

Common neurological disorders

A **neurological disorder** is any disorder in the nervous system, which includes the brain, spine and nerves.

Degenerative myelopathy

Degenerative myelopathy (DM) in dogs is a disease that affects the **spinal cord** of older dogs, although we don't know why. This disease is **degenerative**, so gets worse with time. Dogs will start to wobble, as generally a loss of coordination occurs first in the hind limbs as the white matter in the spinal cord starts to degenerate. Nerves start becoming affected, and so communication becomes lost. There is no treatment, so this should be managed by good husbandry and health management plans.

A dog undergoing a veterinary examination.

Some breeds of dog, such as the German shepherd, are more susceptible to degenerative myelopathy.

White matter

White matter contains the nerve fibres of neurons. So much **myelin** surrounds them that it makes the tissue appear white.

Grey matter contains several components such as the neuronal cell bodies, so it appears darker in colour.

🔗 **Links** For more information on the myelin sheath see page 56.

Listeriosis

A bacterial infection, **listeriosis** can cause encephalitis (swelling of the brain), abortion and blood poisoning. **Symptoms** include disorientation, facial paralysis, constant salivation, collapse and involuntary movement. If untreated it can lead to death.

Seizures

There are many causes of **seizures**, such as brain injury, organ failure, epilepsy and infection. Prior to a seizure, the animal may be nervous, restless, shaking or salivating. During the seizure, the animal may lose consciousness and have convulsions or may hallucinate. Consequences will depend on the cause, duration and type of seizure.

Now try this

Describe the signs that an animal may be about to have a seizure.

Structure and function of blood

Blood transports materials and heat around the body, and helps to protect against disease.

Component	Structure	Function
Red blood cells (erythrocytes)	They have a **biconcave** shape (flattened disc shape) to maximise their surface area for oxygen absorption. They are small and flexible so that they can fit through narrow blood vessels.	The **oxygen** deliverers. They also carry back the waste gases or carbon dioxide. They contain a protein called **haemoglobin**. Red blood cells have **no nucleus**, so they can contain **more** haemoglobin. *Haemoglobin contains iron, which combines with oxygen to give our blood a red colour.*
White blood cells (leukocytes) Lymphocyte Eosinophil Neutrophil Monocyte Basophil	Various structures of white blood cells can be found, including granular cytoplasm, a large nucleus and a lobed nucleus.	**Defend** the body against disease. Consist of **B lymphocytes** responsible for making **antibodies**, and **T lymphocytes** which initiate the **immune response**.
Platelets	Fragments of cells with proteins attached to their surface; these allow them to stick together during clotting.	Make up the rest of the blood. These cells help your body **repair** by stopping bleeding after illness or injury.
Plasma	A clear, pale straw coloured liquid which makes up the fluid component of blood.	Is a liquid part of blood and is involved with material transport such as hormones, carbon dioxide and waste. Plasma makes up just over half of the volume of blood (55%).

 Links There is more information on gas exchange and blood proteins on page 84.

Now try this

What is the function of platelets?

The circulatory system

The circulatory system is also called the **cardiovascular system**, or the vascular system. It allows blood to circulate around the body, and transports nutrients, oxygen, carbon dioxide, hormones, waste and blood cells to and from each cell in the body.

Circulatory system

The circulatory system involves the **circulation of blood** and is important for delivering **oxygen** and **nutrients** and also **removing waste** products. It is made up of the **heart**, **blood** and **blood vessels**. The heart **pumps blood** around the body by contracting and relaxing. Blood is pumped around the heart and body via a network of vessels: **arteries**, **veins** and **capillaries**.

It is a **double** circulatory system. This means the blood travels through two circuits: the pulmonary and systemic.

 Links Double circulatory circuits are discussed further on page 81.

Blood vessels

Arteries: Carry **oxygenated** blood at **higher pressure** from the heart to parts of the body so they have a thick layer of elastic tissue in the wall.

Veins: Carry **deoxygenated** blood at **lower pressure** from the body back to the heart so have less elastic tissue than arteries. Valves stop the blood from flowing backwards.

Capillaries: Have thin walls that allow **exchange** of compounds such as nutrients, glucose, oxygen and carbon dioxide between blood and tissues.

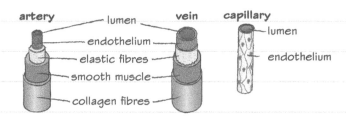

Structure of the heart

Mammals and birds have **four** chambered hearts. This system is very efficient and does not allow **oxygenated** (red) and **deoxygenated** (blue) blood to mix. The **aorta** is the largest artery, which is very elastic and thick as it has to withstand large pressures as it carries oxygenated blood away from the heart. The **vena cava** transporting deoxygenated blood to the heart is made up of the **superior and inferior** vena cava, and is not as thick as it is not under as much pressure.

Heart chambers

There are two types of chambers: **atria** and **ventricles**. Both have muscular walls which helps them pump blood. The ventricles are more muscular than the atria, as the blood must be forced out of the heart at a higher pressure and around the body. The left side of the heart is more muscular as it pumps blood around the body, while the right side just pumps blood to the lungs.

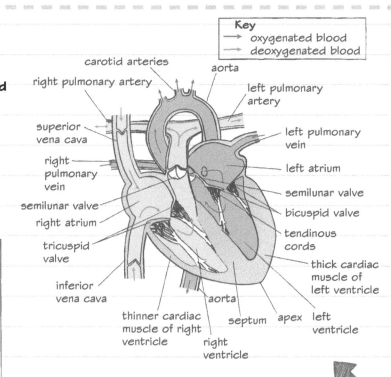

Heart **valves** open and close allowing blood to flow due to pressure changes in the chambers. They are important in **preventing backflow** of blood.

Now try this

1 What do arteries carry?

2 Compare the structure of the aorta and vena cava in relation to their function.

The cardiac cycle

The **cardiac cycle** is the series of events that occurs when the heart beats. This circulates blood through the **pulmonary** and **systemic** circuits that make up the double circulatory system.

Double circulatory system

A double circulatory system is more efficient in transporting oxygenated blood around the body because the oxygenated and deoxygenated blood are kept separate. The **pulmonary** circuit carries blood **to the lungs** to be **oxygenated** and then **back to the heart**. In the lungs, carbon dioxide is removed from the blood, and oxygen taken up by the haemoglobin in the red blood cells.

The **systemic** circuit carries blood **around the body** to deliver the oxygen and returns **de-oxygenated** blood to the heart.	This is broken down further into the **cardiac cycle**, which is the sequence of mechanical and electrical events that are repeated every heartbeat. It includes two phases:	Diastole – relaxation phase, during this stage blood pressure is at its lowest.	Systole – contraction phase, this is when the blood pressure rises.	The frequency of the cardiac cycle is monitored by the heart rate, which is recorded by number of beats per minute.

The cardiac cycle in the left side of the heart

Blood drains into left atrium from lungs along the pulmonary vein.	Raising of the blood pressure in the left atrium forces the left tricuspid valve open.	Contraction of the left atrial muscle (left atrial systole) forces more blood through the valve.	As soon as left atrial **systole** (muscle contraction) is over, the left ventricular muscles start to contract. This is called left ventricular systole.	This forces the left tricuspid valve to close and opens the valve in the mouth of the aorta (semilunar valve). Blood then leaves the left ventricle along the aorta.

Cardiac diastole is when the heart refills with blood. Ventricular diastole is when the ventricles are refilling and relaxed.

The same steps are repeated on the right side at the same time.

The heartbeat

For the heart to beat, **electrical signals trigger** muscles to contract and relax. There is a specific pathway to allow this to happen.

1 **Sinoatrial node** – this is also known as the natural pacemaker as it causes an impulse to travel through the atria causing them to contract and force blood into the ventricles. This node also sets the heart's rhythm and rate.

2 **Atrioventricular node** – this detects the impulse travelling through the atria and redirects the impulse to the bundle of His. This causes a delay, slowing the spread of the electrical impulse across the heart and allowing the atria to contract before the ventricles.

3 **Bundle of His** – also called the atrioventricular bundle, this is a group of fibres in the septum which the impulse travels through to the base of the ventricles.

4 **Purkinje fibres** – these fibres act like neurons and are found in the walls of the ventricles. The impulse from the bundle of His reaches the Purkinje fibres which cause the ventricles to contract.

Baroreceptors

Stretch receptors (baroreceptors) in the heart detect changes in the pressure of blood filling the atria. These send signals to the CNS which triggers **vasodilation**. This reduces the pressure in the blood.

Now try this

1 Name the main artery leading from the heart.

2 What is known as the 'natural pacemaker', setting heart rate and rhythm?

The respiratory system

The **respiratory system** allows air to pass in and out of an animal's body for efficient **gas exchange** to happen.

Mammalian respiratory system

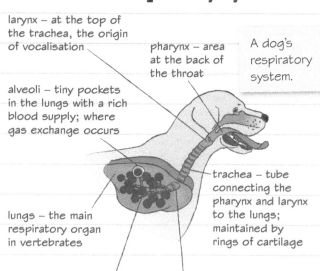

larynx – at the top of the trachea, the origin of vocalisation

pharynx – area at the back of the throat

A dog's respiratory system.

alveoli – tiny pockets in the lungs with a rich blood supply; where gas exchange occurs

trachea – tube connecting the pharynx and larynx to the lungs; maintained by rings of cartilage

lungs – the main respiratory organ in vertebrates

bronchioles – smaller airways connecting the bronchi to the alveoli

bronchi – main airways leading in to the lungs; one for each lung

 Links For gas exchange, see page 84.

Avian respiratory system

The avian respiratory system has many differences to the mammalian system.

- **Nares:** the avian equivalent of nostrils
- **Larynx:** not used for vocalisation; the **syrinx** is used for vocalisation instead
- 7 or 9 **air sacs** which extend into some bones
- **No diaphragm:** pressure changes move air through the air sacs
- **Air capillaries** instead of alveoli for gas exchange.

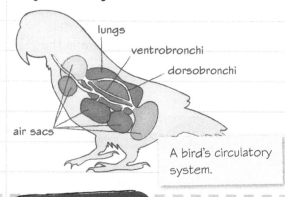

lungs
ventrobronchi
dorsobronchi
air sacs

A bird's circulatory system.

Mammalian ventilation (breathing)

When breathing, the ribs, intercostal muscles and diaphragm all play important roles.

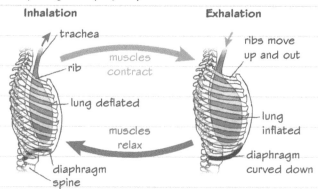

Inhalation

trachea
muscles contract
rib
lung deflated
muscles relax
diaphragm
spine

Exhalation

ribs move up and out
lung inflated
diaphragm curved down

	Inhale	Exhale
Diaphragm and intercostal muscles	Contract	Relax
Thorax volume	Increase	Decrease
Pressure	Decrease	Increase
Air flow	In	Out

Air flow in the body

atmosphere ▶ trachea ▶ bronchi ▶ bronchioles ▶ alveoli ▶ blood ▶ cells ▶ blood ▶ alveoli ▶ bronchioles ▶ bronchi ▶ trachea ▶ atmosphere

Avian ventilation

Birds need two respiratory cycles to move the air through the entire respiratory system. Mammals only need one cycle.

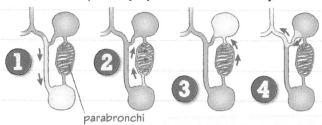

parabronchi

Air movement in bird respiration.

1 **Inhalation:** air passes through the larynx, trachea and into the posterior air sacs

2 **Exhalation:** air moves from the posterior air sacs to the lungs via the ventrobronchi and dorsalbronchi

3 **Inhalation:** air moves from the lungs to the cranial air sacs

4 **Exhalation:** air moves from the cranial air sacs through the syrinx, trachea and out of the nares

Now try this

Where does gas exchange take place?

Respiration

Respiration is the release of energy from glucose, or another organic chemical such as a lipid or protein. This **energy** is used for growth, making new materials, movement and (in some animals) heat production. Respiration is **not** breathing.

Aerobic respiration

Aerobic respiration is the release of oxygen from the breakdown of glucose by combining with oxygen. Oxygen combines with glucose to produce energy. This happens in the mitochondria of cells.

 Links   Cell structure is discussed on page 46.

It is 18–19 times more efficient than anaerobic respiration:

glucose + oxygen → carbon dioxide + water (+ energy)

> Remember that energy is stored in ATP, which is the 'universal energy currency'.

Anaerobic respiration

Anaerobic respiration means respiration **without air**. It is a short-term energy production method, used when animals cannot exchange enough oxygen to carry out aerobic respiration, such as during strenuous exercise. It happens in the cytoplasm of cells. Anaerobic respiration is **not** as efficient as aerobic respiration and it also leaves a poisonous chemical, **lactic acid**. This stops muscles working well and causes pain:

glucose (broken down to) → lactic acid (+ energy)

Homeostasis and respiration

The heart rate, respiratory rate and volume play a large role in **homeostasis** for blood oxygen and carbon dioxide levels. Receptors in the blood vessels detect changes in blood composition and initiate changes in the respiratory and circulatory systems to keep conditions within safe limits.

	Blood oxygen increase	Blood carbon dioxide increase
Heart rate	Decreases	Increases
Respiratory rate	Decreases	Increases
Respiratory volume	Decreases	Increases

Oxygen debt

With a big increase in the level of exercise being done, the body may not have enough oxygen to respire aerobically. Instead, **anaerobic respiration** can occur and lactic acid builds up in muscles. Lactic acid can damage cells and tissues, but can be broken down using oxygen.

Because of the temporary shortage of oxygen, an '**oxygen debt**' occurs. After anaerobic respiration has stopped, an increase in ventilation rate and deeper breathing or panting provides the oxygen to 'pay back' the lactic acid build-up. This continues until the extra lactic acid has been broken down.

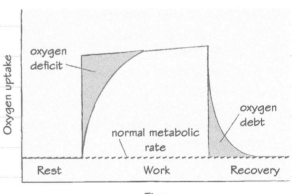

Now try this

You are working at a horse racing yard and find that one of the horses is suffering from muscle soreness.

What could cause this and why?

Gas exchange and blood proteins

Gas exchange occurs in the lungs in tiny sacs known as **alveoli**. These are surrounded by a network of capillaries which are vital for gas exchange to take place.

Gas exchange at the lungs

The gases move by **diffusion**, from where they have a high concentration to where they have a low concentration.

- **Oxygen diffuses** from the air in the alveoli into the blood.
- **Carbon dioxide diffuses** from the blood into the air in the alveoli.

Alveoli in the lungs are adapted for efficient **gas exchange**. They are very **thin** with a large **surface area**.

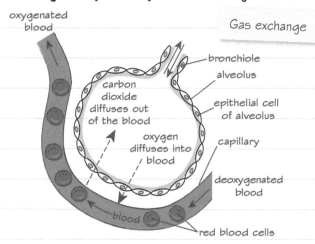

Gas exchange

How it all fits together

Haemoglobin in erythrocytes (red blood cells):

- loads oxygen at the lungs and unloads it at respiring tissues
- loads carbon dioxide at respiring tissues and unloads it at the lungs.

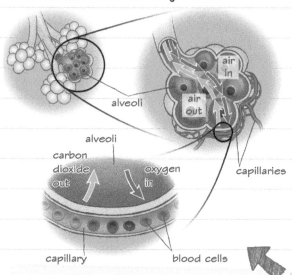

Inhaled air contains 21% oxygen and 0.04% carbon dioxide, while exhaled air contains 16% oxygen and 4% carbon dioxide.

Blood proteins and oxygen

Haemoglobin	Myoglobin	Foetal haemoglobin
<ul><li>Protein found in red blood cells</li><li>Carries four oxygen molecules</li><li>Used to transport oxygen all over the body</li><li>Removes carbon dioxide as a waste product</li><li>**Oxyhaemoglobin** is the oxygenated form of haemoglobin</li></ul>	<ul><li>Found in muscle cells</li><li>Is an oxygen and iron binding protein which releases oxygen to muscle cells when oxygen concentration is low</li><li>Only carries one oxygen molecule</li><li>Acts as an oxygen store</li><li>**Oxymyoglobin** is the oxygenated form of myoglobin</li></ul>	<ul><li>A foetus cannot use its lungs, and so relies on exchanging oxygen and carbon dioxide via diffusion from its mother's blood</li><li>**Foetal haemoglobin** has a **higher affinity** for oxygen, and so is more efficient at extracting oxygen from maternal circulation</li></ul>

Bohr effect

When rates of **respiration** increase, there is an increased amount of carbon dioxide in the blood which dissolves and forms carbonic acid. This lowers the pH of the blood, and so causes the haemoglobin in the blood to release more oxygen. Because of this, when exercising, the muscles are able to continue to work at the same rate.

Now try this

1. What features of the alveoli improve the efficiency of gas exchange?
2. What are the **two** main gases that are being exchanged between the lungs and the circulatory system?

The lymphatic system

The **lymphatic system** is very important in helping the body **fight infection**. Not all animals have a lymphatic system, but you need to know its composition and actions.

The lymphatic system

The **lymphatic system** works alongside the blood circulatory system to help get rid of **toxins** and any **unwanted material**. It is made of a network of lymph vessels, tissues and organs.

Lymph is a clear fluid rich in white blood cells especially lymphocytes. Lymph is formed from blood plasma, which is forced out of blood vessels by pressure and moves around tissues until it collects in the lymph vessels.

Lymph organs – the thymus and spleen are both directly connected to the lymphatic system.

Lymph nodes, e.g. axillary node – found along lymph vessels, these filter the lymph to get rid of toxins, waste and pathogens.

Lymph vessels – act like veins and capillaries to transport the lymph.

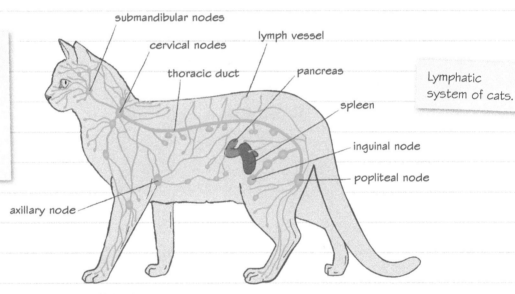

Spleen – controls the levels of erythrocytes in the blood and is where B-cells mature; these secrete antibodies.

submandibular nodes
cervical nodes
thoracic duct
lymph vessel
pancreas
spleen
inguinal node
popliteal node
axillary node

Lymphatic system of cats.

Lymph composition varies depending on location in the body, health of the animal and diet. In general, it contains: protein, glucose, lipids, water, leukocytes, salt.

Thymus – located in front of the heart, this is where T-cells mature; these destroy virus-infected cells and can instruct the body to release other cells to target the infection.

Functions of the lymphatic system

- ✓ Returns interstitial fluid from tissues to the circulatory system.
- ✓ Filters lymph to remove waste and toxic materials.
- ✓ Produces and transports antibodies and lymphocytes.
- ✓ Transports digested fat from the intestine to other sites in the body for storage.

Now try this

1 Where do T-cells mature?
2 What cells produce antibodies?

Circulatory system disorders

A circulatory disorder is any disorder or condition that affects the circulatory system. Circulatory disorders can arise from problems with the heart, blood vessels or the blood.

Mitral valve (bicuspid valve) disease (MVD)

Valves help regulate the flow of blood and also prevent backflow. MVD causes the **mitral valve** to **degenerate**, so it does not close fully with each pumping action of the heart and causes blood to **flow backwards**. This eventually worsens and the **valve collapses**, causing heart failure. **Symptoms** include a heart murmur, shortness of breath, coughing and a reduced exercise ability.

Treatment relies on the use of drugs to remove excess fluid and to ease the burden on the heart. A low-sodium diet and exercise restriction can also aid the treatment of MVD.

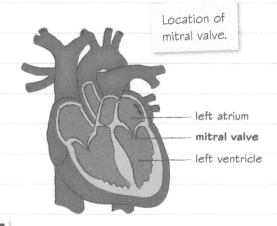

Location of mitral valve.

left atrium
mitral valve
left ventricle

Mitral valve disease is very common in Cavalier King Charles Spaniels and it inevitably leads to **heart failure**.

Heart murmurs

Heart murmurs are caused by **abnormal blood flow** through the heart from an abnormal heart structure (problems with valves or chambers). **Symptoms** vary depending on the extent of the structural problem, but coughing, weakness and problems exercising are common. The **prognosis** varies depending on the cause of the murmur; some heart murmurs can resolve themselves, others will be treated with diet, medication and possibly surgery.

Von Willebrand disease

Von Willebrand disease is caused by insufficient amounts of a **clotting factor**, resulting in blood not clotting properly. There are three types, with **Type 1** being the least severe and **Type 3** the most severe. Symptoms include unexplained bleeding, blood in urine or faeces, excessive bleeding from minor wounds or surgeries, and lameness in joints. No treatments are available for von Willebrand disease. Management involves reducing the potential for bleeding to occur and managing bleeds that do occur.

Now try this

Cavalier King Charles Spaniels are prone to mitral valve disease and it is recommended that they receive cardiac ultrasounds before entering a breeding program.

Why do you think this is?

Male reproductive system in mammals

Although there are differences between species, there are general structures within the mammalian reproductive system. You will need to be able to identify these and their functions.

Structure of male reproductive system

The function of the **male reproductive system** is to produce and store **sperm**, and to discharge sperm into the female and **fertilise** the **egg**.

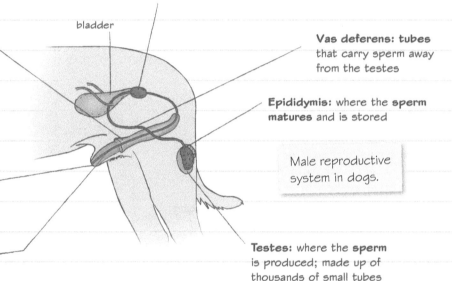

Prostate gland: adds secretions to the sperm as it passes along the vas deferens, which contain fructose and enzymes; combined with sperm, these fluids form semen, which carries sperm into the female reproductive tract where ova can be fertilised

Bulbus glandis: also called a knot, is an erectile tissue structure on the penis of canid mammals; during copulation the tissue becomes enlarged and locks the male to the female.

bladder

Vas deferens: tubes that carry sperm away from the testes

Epididymis: where the **sperm matures** and is stored

Male reproductive system in dogs.

Penis: located on the **outside** of the body, semen and urine can pass though.

Urethra: tube that **carries** sperm, semen and urine

Testes: where the **sperm** is produced; made up of thousands of small tubes

Differences between species

There is much variation in penis structure between different species. These correspond to the shape of the female's vagina.

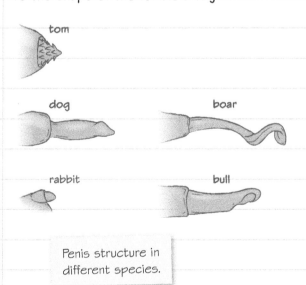

tom

dog   boar

rabbit   bull

Penis structure in different species.

Internal structure of the penis

The penis contains three columns of erectile tissue.

Os penis: a **bone** which is present in some species such as cats and dogs to maintain an erection.

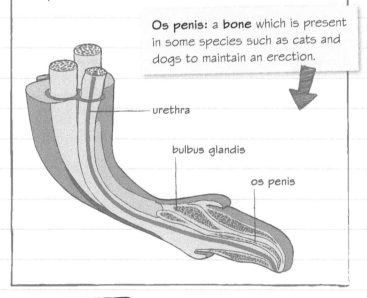

urethra

bulbus glandis

os penis

Now try this

Describe the function of the epididymis.

Female reproductive system in mammals

The **female reproductive system** is designed to **make eggs** and provide the perfect environment for **fertilisation**. It also protects and looks after the **foetus** until birth.

Female reproductive system

Ovaries: where the **egg** and the **hormones** oestrogen and progesterone are made.

Oviduct (fallopian tube): the egg released from the ovary is passed into the oviduct, where fertilisation happens; the fertilised egg begins to travel down the oviduct and, when it reaches the uterus, it implants in the uterus lining and pregnancy is established.

Uterus: thick muscular structure where fertilised **egg implants**; stretches to accommodate growth of the foetus, and later contracts to expel the foetus at birth; made up of the **uterine horns** and uterine body; the shape varies with **species**.

Cervix: a muscular **barrier** to seal the uterus; provides **protection** for the reproductive system from **infection**.

Vagina: shaped to fit the penis during mating; produces **acidic mucus** to kill bacteria and **prevent** infection.

Vulva: external part of female reproductive system; in some species the vulva swells up and is a visual sign of **oestrous** which shows the male they are ready to mate.

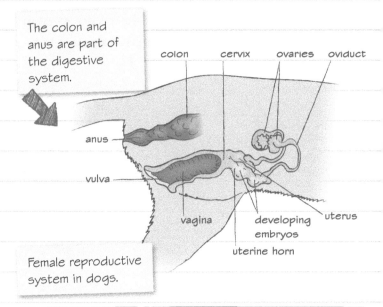

The colon and anus are part of the digestive system.

Female reproductive system in dogs.

colon   cervix   ovaries   oviduct

anus

vulva

vagina   developing embryos   uterus

uterine horn

The placenta

Once pregnancy is established, the **placenta** provides a connection between the foetus and maternal tissue to provide nutrition, respiration, digestion, protection and excretion. The placenta is a temporary organ which develops along with the foetus, and is attached to the uterus. It contains both foetal (from the umbilical cord) and maternal (from the uterine wall) blood vessels.

Uterus differences

The uterus consists of **three** layers. The outer layer is the **perimetrium** which supports the shape of the uterus, the **myometrium** is the muscular layer which contracts and the **endometrium** forms the inner layer and is where implantation occurs. The shape of the uterus differs between species.

Cats and dogs have a similar uterus structure but differ in size of the uterus.

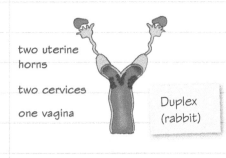

two uterine horns

two cervices

one vagina

Duplex (rabbit)

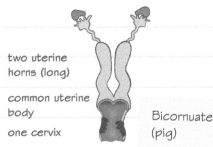

two uterine horns (long)

common uterine body

one cervix

Bicornuate (pig)

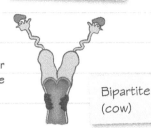

smaller uterine horns

Bipartite (cow)

Now try this

1 What is the function of: (a) the cervix (b) the uterine?
2 What type of uterine horn does a pig have?

Fertilisation and gestation length in mammals

The process from fertilisation to implantation is the same for most mammals. There can be big differences in how long this process takes between species.

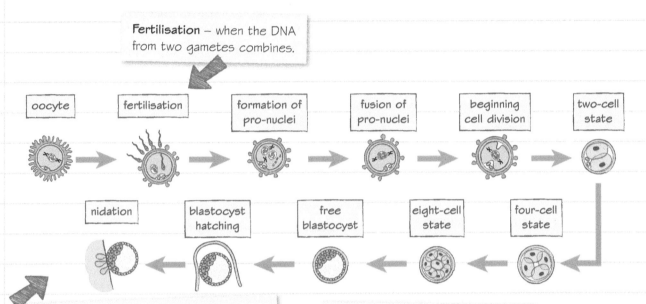

Fertilisation – when the DNA from two gametes combines.

oocyte | fertilisation | formation of pro-nuclei | fusion of pro-nuclei | beginning cell division | two-cell state

nidation | blastocyst hatching | free blastocyst | eight-cell state | four-cell state

Implantation – when a fertilised egg attaches itself to the lining of the uterus.

Gestation – the length of time it takes from conception to birth.

Gestation length

Species	Average gestation
Cat	61–72 days
Cow	279–287 days
Dog	58–70 days
Pig	114–115 days
Rabbit	30–32 days

Pregnant cow

Delays in implantation/nidation

In some species, there is a time delay between fertilisation and implantation. This is because the fertilised egg needs to grow to be able to join on to the uterus lining without embedding in it; called **centric implantation**.

This time delay depends on the species.

Species	Time delay between fertilisation and implantation
Rabbit	6.5 days
Cat	10–12 days
Dog	6–10 days
Pig	12–14 days
Cattle	30 days

Now try this

A farmer has arranged for his sow to be artificially inseminated. After how many days should he expect the litter to be born?

Reproductive hormones

Animal reproductive systems are controlled by **hormones**, which are released by **glands** around the body.

Gametogenesis

The **male gamete** is called the **sperm** and the **female gamete** is known as an **ovum**.

The production of **sperm** is known as **spermatogenesis** and this is a **continuous** process in the body. The production of an **ovum** is known as **oogenesis** and this is part of the **menstrual cycle**.

These processes are known as **gametogenesis**, and are controlled by two very important hormones known as **luteinising hormone (LH)** and **follicle stimulating hormone (FSH)**.

Spermatogenesis

This process occurs **continuously** in the body:

☑ LH stimulates testosterone to be produced.

☑ This causes the production and maturation of **sperm**.

☑ **Inhibin** is produced by Sertoli cells and **inhibits FSH** to stop spermatogenesis.

Oogenesis

This is part of the **menstrual cycle**:

☑ **FSH** stimulates one or more follicles to mature.

☑ **LH** stimulates mature follicle to produce more **oestrogen** and **inhibit LH** and **FSH** production.

☑ **Oestrogen** (+ inhibin) from mature follicle **inhibits** secretion of **FSH** so no more follicles mature.

☑ However, an increase of **oestrogen** levels stimulate **LH** and **FSH** production which leads to **ovulation**.

Effect of other hormones

The reproductive system is also affected by a variety of other hormones within the body.

Hormone	Role and effects
Gonadotropin releasing hormone (GnRH)	Hormone **secreted** by the **hypothalamus** and **controls** other reproductive hormones. **Light** stimulates the **pineal gland** which **reduces melatonin** secretion, thus allowing **GnRH** to be secreted which stimulates the production of **LH and FSH**.
Oestrogen	Hormone that **rebuilds the uterine lining** (endometrium) after **menstruation**. Stimulates the release of **LH** by the pituitary.
Progesterone	Maintains the **lining** of the **uterus** ready for a fertilised egg. It **inhibits the production of FSH** by the pituitary. Progesterone is mostly produced by the **corpus luteum**, and in smaller amounts by the ovaries.
Testosterone	Produced by the **testes**, this enables the development of sex organ tissues and secondary sexual characteristics.
Androgen-binding hormone	Produced by the **seminiferous tubules**, this helps to maintain high concentrations of testosterone in the testes.
Prostaglandins	Help with uterine contractions; these are produced by most cells.
Oxytocin	Produced by the **hypothalamus** and released by the pituitary gland. It helps stimulate uterine contractions and lactation.
Cortisol	Generally considered the **stress hormone**, cortisol is linked to increased maternal interest in their offspring.

The corpus luteum is the sac of cells remaining after the ova has been released during ovulation.

Now try this

1 What is the production of sperm known as?
2 Name the hormone important in rebuilding the uterine lining.

The oestrous cycle

When female mammals become sexually mature, their reproductive systems begin to undergo a **cycle** of changes.

Different oestrous cycles

The **oestrous cycle** can vary depending on the animal, for example:

- **Polyoestrous** animals – become fertile more than once per year, e.g. cattle, pigs, mice.
- **Seasonally polyoestrous** animals – have more than one oestrous cycle, but only during certain periods of the year, e.g. horses, sheep, goats, deer.
- **Monooestrous animals** – have one oestrous cycle every year, e.g. dogs, wolves, foxes, bears.

Polyoestrous in sheep

Seasonal polyoestrous in sheep is linked to high light levels causing **melatonin** to be released by the **pineal gland**. Melatonin inhibits the release of LH and FSH. When light levels start to decrease, the inhibition of LH and FSH decreases, and the presence of increasing levels of LH and FSH triggers **oestrous**.

Stages of the oestrous cycle

In animals, there are **four main stages** to the oestrous cycle.

Dioestrous – levels of progesterone peak.

Pro-oestrous – follicle enlarges, decrease in progesterone levels leads to an increase in oestrogen levels; GnRH is released.

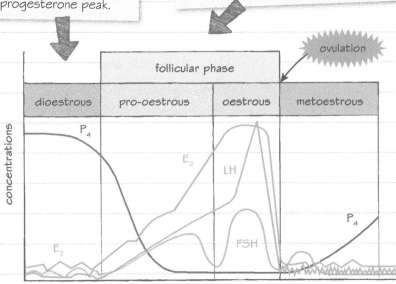

Oestrous (heat) – oestrogen increases along with LH and FSH, leading to ovulation.

Metoestrous – the corpus luteum starts to decay which increases progesterone levels. This inhibits oestrogen release and the level drops.

Links For further information regarding other hormones that are involved in the reproductive system see page 92.

Now try this

Describe how the level of oestrogen changes during the pro-oestrous and oestrous phases of the cycle.

Roles of hormones in parturition and lactation

Hormone levels change during parturition and lactation. You will need to know the roles of the different hormones during these stages.

Role of hormones

Progesterone levels decrease allowing other hormones to effect uterine muscles. In some species, this drop in progesterone depends on the placenta, in others the corpus luteum.

↓

Oestrogen and oxytocin levels rise leading to uterine contractions and cervical softening. Parturition begins.

↓

Relaxin, secreted by the ovaries, relaxes the pelvic ligaments and opens the birth canal.

↓

Oestrogen and progesterone levels drop allowing prolactin to be secreted, triggering lactation.

Giving birth

Parturition is also known as **giving birth**. This is called lambing in sheep, kidding in goats, whelping in dogs and calving in cattle.

Links Parturition is also discussed on page 22.

Cortisol

When animals are stressed their body releases cortisol, which is similar in structure to progesterone. The body prioritises making cortisol, so progesterone levels drop which causes a high oestrogen: progesterone ratio and can lead to early labour.

Links Gestation length for various species can be found on page 89.

Lactation

Lactation is the production of milk by the female's **mammary system**. It allows a mother to provide her offspring with highly **nutritious milk** to aid its survival. The offspring suckle on the mammary system, which stimulates milk release. It is vital that the offspring consume milk as it supports them until they are weaned and ready to move onto other foodstuff.

Hormones are necessary to start **milk production**, stimulate **milk let-down** and then **restart** the **reproductive** process.

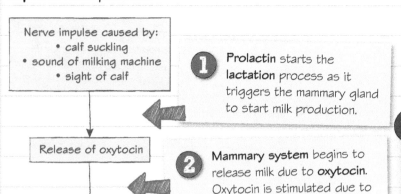

Nerve impulse caused by:
- calf suckling
- sound of milking machine
- sight of calf

↓

Release of oxytocin

↓

Release of milk

1 Prolactin starts the lactation process as it triggers the mammary gland to start milk production.

2 Mammary system begins to release milk due to **oxytocin**. Oxytocin is stimulated due to the offspring **suckling** at the teat (or milking).

Now try this

1 What hormone is responsible for uterine contractions during parturition?

2 Describe the milk let-down process.

Reproductive system in birds

The **reproductive system** of **birds** is different to that of other animal species.

The female reproductive system

The **ovary**, **oviduct** and the **cloaca** are the functional parts to the female poultry reproductive system. Before a female reaches sexual maturity, the right ovary and oviduct **degenerate** leaving **one** functional **ovary**.

The **ovary** appears as a cluster, which is composed of oocytes. Here, ovum (eggs) may develop over time. **Ovum** formation occurs from the collection of **lipid** particles from the blood which form the **yolk**.

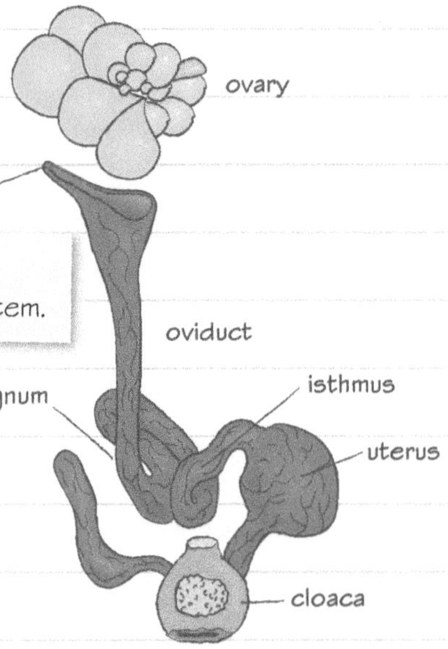

The **oviduct** consists of five parts:

- **Infundibulum** – receives egg and is where fertilisation takes place.
- **Magnum** – secretes the albumen (the white).
- **Isthmus** – adds the shell membrane.
- **Uterus** – secretes the shell and shell pigment.
- **Cloaca** – where the egg and waste products (digestive and urinary) pass out.

The female avian reproductive system.

(Diagram labels: ovary, infundibulum, oviduct, magnum, isthmus, uterus, cloaca)

The male reproductive system

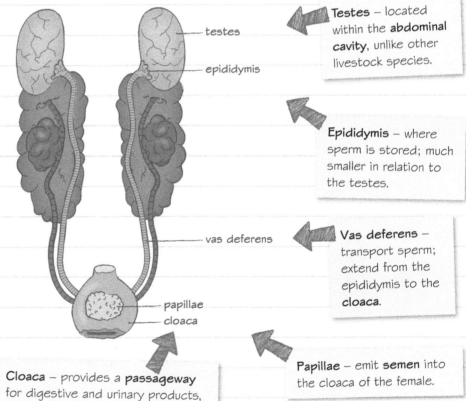

Androgens are the male sex hormones such as testosterone, which are produced by the testes and are responsible for the production of **sperm** and leading **sexual activity**. They control secondary sexual characteristics such as comb length and influence 'pecking order'.

Phallus – a **copulatory organ**, which becomes inflamed with lymph during mating, and allows semen to be deposited into the female.

(Diagram labels: testes, epididymis, vas deferens, papillae, cloaca)

Testes – located within the **abdominal cavity**, unlike other livestock species.

Epididymis – where sperm is stored; much smaller in relation to the testes.

Vas deferens – transport sperm; extend from the epididymis to the **cloaca**.

Papillae – emit **semen** into the cloaca of the female.

Cloaca – provides a **passageway** for digestive and urinary products, but also reproductive tracts.

1 State how many ovaries an adult female chicken has.

You have completed a dissection on the female reproductive system of a chicken. You have identified the oviduct.

2 Name the parts of the oviduct.

Structure of an avian egg

Eggs provide the environment and nutrients developing chicks require. You will need to know about the **structure** of an egg and the **function** of the main components.

Shell – this is made of calcium carbonate and is hard to protect the developing chick. It's semi-permeable which allows for gas exchange. It also stops unwanted substances like toxins from entering the shell.

Inner and outer shell membranes – these are made up of proteins and lie between the eggshell and albumen. They are an effective defence against invading bacteria.

Yolk – this is the yellow part of the egg. The yolk will provide food and nutrition as the embryo forms.

Chalaza – this is the spiral strand that holds the yolk in the egg white. There are two of these, and together they are known as chalazae.

Albumen – this is also known as the 'egg white' and contains a variety of different proteins.

Air cell – this is a gap located between the shell membranes.

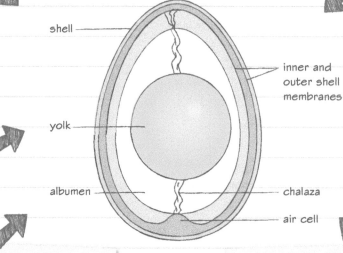

shell — inner and outer shell membranes

yolk

albumen — chalaza

— air cell

Different species

Eggs come in many different shapes and sizes. This can depend on many factors such as breed and species, but the principles of development are the same for all eggs.

A quail egg, hen egg, duck egg and a goose egg.

Extra-embryonic membranes

During development, the embryo receives protection and nourishment from the four extra-embryonic membranes:

1. **Chorion** surrounds the embryo and allows for gas exchange.

2. **Amnion** forms a sac which helps to protect the embryo and help hold the amniotic fluid which acts as a shock absorber.

3. **Allantois** develops a large circulatory system which allows for respiratory, excretory and digestive functions.

4. **Yolk sac** has a digestive function as is used as a food source for the embryo, but is also used by the chick once hatched.

Now try this

Chicken embryos take approximately 21 days to develop and hatch once fertilised.

Describe **two** features of an egg that help protect the embryo during its development.

Try to describe one feature from the diagram at the top of the page and one extra-embryonic membrane.

Egg formation

Egg formation for egg-laying animals is more complicated than in live-bearing animals. You need to know the different **stages in egg formation** for egg-laying animals.

Fertilisation

This process starts with copulation, the **male** placing his **sperm** into the **oviduct** of the **female** by depositing sperm from the **papillae** to the wall of the female's **cloaca**. The sperm can linger in the oviduct for up to **three weeks**. However, they only have full fertilising ability for approximately **six days**.

The **oviduct** is where the egg is developed before it is laid. An egg is **formed** over a period of approximately **25 hours**.

> **Links** See page 93 for bird reproductive structure.

Structure of the oviduct

The egg yolk (**ovum**) is **released** from the **ovary** to the **infundibulum** (funnel) **of the oviduct** (remember there is only one functional oviduct). The oviduct has **five parts** which play separate roles in **producing an egg**.

Part of the oviduct	How long egg spends here	Functions of the parts in egg formation
Infundibulum (funnel)	15 minutes	Here, the yolk is received from the ovary and is fertilised by the sperm (sperm is not always present – commercial eggs are not fertilised)
Magnum	3 hours	The albumen (white) is deposited and layered around the yolk
Isthmus	1 hour	Both inner and outer shell membranes are added
Uterus (shell gland)	21 hours	Water is added, which makes the outer white thinned; shell material is added, which is made of calcium carbonate; some pigments are added to the shell
Cloaca	Less than a minute	The egg is laid

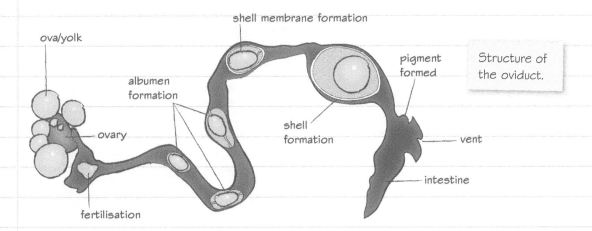

Structure of the oviduct.

Now try this

Where is the site of fertilisation in a chicken?

Embryo development

The **incubation period** of eggs differs between bird species, and there are several factors that can affect the production of eggs.

Incubation period

A female chicken can lay approximately 200–300 eggs a year. They usually lay an egg every 24 hours, however, factors such as incubating and moulting cause no eggs to be laid.

Incubation is the period in which the embryo is developed inside the egg, until it hatches.

Incubation periods

Species	Incubation period
Chickens	21 days
Pheasant	23–24 days
Ducks	28 days
Barn owls	31 days

Factors affecting egg production

Different factors affect egg production:

- **Availability of light** (photoperiod) – egg production is **stimulated by light**, so in the middle of winter egg production decreases dramatically and can even stop.
- **Food availability**, **diet** and **health** – poor nutrition and illness can cause a cease in egg production.

- **Temperature** – optimum temperatures for egg laying are around 11–26 °C. Below and above this, the rate and quality of egg production decreases.
- **Other factors** – such as type of breed, housing, husbandry, moulting, age and weather can affect egg production.

In **commercial** chicken egg production systems, the use of artificial lighting helps 'trick' the chickens into remaining in **production**. This can be achieved by increasing light hours all year round.

Incubation

It is important to incubate the eggs at the right **temperature** and **humidity**. Hens may sit on their eggs, while commercial production systems utilise incubators.

- In commercial systems, temperature varies from **37–39 °C**. Underheating and overheating eggs can lead to abnormal embryos.
- **Relative humidity level** is kept at approximately 60% for the first 18 days and 70% for the last three days. This is to stop the egg drying out.
- **Ventilation** is important as **oxygen** enters through the shell. This is increased gradually as the embryo begins to hatch.
- **Turning the eggs** stops the developing embryo sticking to the inside of the shell. This occurs **at least three times** a day.

Eggs in an incubator.

Now try this

1 How long is the incubation period of a chicken egg?

> You are working for a poultry organisation and have been asked to produce a leaflet to inform hobby chicken breeders on egg production.

2 Identify **four** factors that can influence egg production in chickens.

The excretory system

Excretion is a very important process which allows **waste** products to be **removed**. It also helps to maintain a constant internal environment and this process is carried out by the excretory system.

The excretory system and its structure

The excretory system consists of the kidneys, bladder, ureters and urethra. Its purpose is to filter the blood to remove waste products which would be toxic if they built up.

  In mammals:

- the **ureters** are two **tubes** which carry urine from the **kidneys** to the **bladder**.
- the **bladder** stores the urine before it is removed from the body.
- the **urethra** allows the passing of urine from the bladder and out of the body.

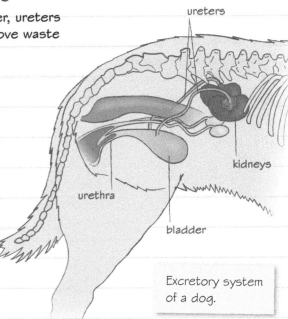

Excretory system of a dog.

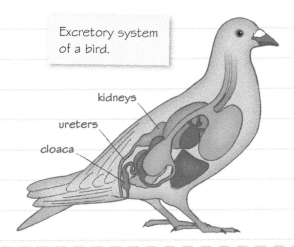

Excretory system of a bird.

② **In birds:**

- the **ureters** are two **tubes** which carry urine from the **kidneys** to the **cloaca**.
- the **cloaca** allows urine and other waste products to be cleared from the body.

Nitrogenous waste removal

The excretory system removes nitrogenous **waste products** such as **ammonia**, as well as **salt** and **water**. Nitrogenous waste products come from the breakdown of proteins.

✓ **Mammals** remove ammonia in the form of **urea**; this requires water to allow it to be excreted from the body as **urine**.

✓ **Birds** remove ammonia in the form of **uric acid**; this does not require water and is an insoluble substance.

The kidneys

The **kidneys** are the core organ involved in **waste removal**. They contain millions of **nephrons** – these are where **ultrafiltration** of the blood takes place and urine is produced.

Kidney failure can have devastating effects due to a build-up of toxic products in body tissues.

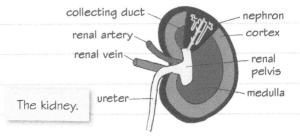

The kidney.

Now try this

1 Explain what the urinary system does.
2 State where urine is stored before being removed from the body in mammals.

Nephrons

Nephrons are an important part of the kidneys.

Structure of nephrons

The **glomerulus** is a 'knot' of capillaries.

1 The **renal (Bowman's) capsule** collects the filtrate from the glomerulus.

2 **Tubules** allow transportation of **filtrate** to and from the **loop of Henle**.

3 **Collecting duct** collects urine and passes to the ureter.

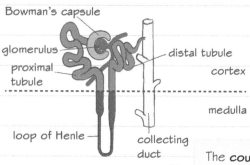

Bowman's capsule

glomerulus

proximal tubule

distal tubule

cortex

medulla

loop of Henle

collecting duct

Nephrons

Nephrons are the site of the three processes of the kidneys:

1 **Ultrafiltration:** removing substances from the blood.

2 **Selective reabsorption:** reabsorbing the substances the body needs.

3 **Urine formation:** substances which are not reabsorbed travel as urine along the nephron to the collecting ducts where the waste travels to the bladder.

The body's water content is controlled by the kidneys via a process known as **osmoregulation.**

Links See page 99 for more on osmoregulation.

The **counter current multiplier mechanism** allows **concentrated urine** to be produced. Animals with longer loops of Henle, e.g. desert species, are able to produce more concentrated urine.

Ultrafiltration

Ultrafiltration occurs in the **renal capsule**. Blood first enters via the **renal artery** and flows through many **capillaries under high pressure** (renal capsule). Specialised cells called **podocytes** make up the basement membrane, leaving small gaps. Here, **small molecules** and **ions** are squeezed out of the blood by the pressure and into the nephron. **Larger** molecules such as **proteins** stay in the blood.

Selective reabsorption

Most of the reabsorption occurs in the proximal convoluted tubule where simple ions, amino acids and glucose, diffuse into the blood via channel proteins. Some larger molecules may be reabsorbed by **endocytosis**. The **loop of Henle** is important in the reabsorption of water and salt.

Anti-diuretic hormone (**ADH**) increases the amount of water reabsorbed in the kidneys. It is released from the pituitary gland to maintain homeostasis.

Counter current multiplier in the loop of Henle

1 **Descending limb** – this is permeable to water. Water leaves the nephron by osmosis and is reabsorbed. As much water as the body needs is reabsorbed here.

2 **Ascending limb** – this is permeable to ions. Because the tissue around the loop has a high concentration of water, Na^+ (sodium) and Cl^- (chlorine) ions diffuse out of the nephron. The concentration of salt in the ascending limb decreases further up the limb. Towards the top of the limb, some Na^+ and Cl^- are actively pumped out of the nephron.

3 **Distal convoluted tubule** – this is where final reabsorption of ions and water occurs to adjust the water balance. ADH has an effect on the amount of water reabsorbed here.

Now try this

Kidneys filter the blood to remove waste.

State the parts of the kidney where the following occur:

1 ultrafiltration   **2** water reabsorption   **3** salt reabsorption.

Osmoregulation

Osmoregulation is maintaining the **balance of salt and water concentrations** in the body to keep a constant osmotic pressure. You will need to know about the mechanisms involved in this.

Mechanism of osmoregulation

Osmoregulation is a physiological process that an organism uses to maintain water balance, through the use of osmoreceptors and the endocrine system.

The **hypothalamus** detects a change in water content via **osmoreceptors** and controls regulation of water levels in the body.

1 **High water content:** causes the body to be hydrated. This is detected by the osmoreceptors in the hypothalamus which send a response to the pituitary gland to secrete less ADH (anti-diuretic hormone). Less water is reabsorbed by the kidneys resulting in more urine being produced so more water is lost.

2 **Low water content:** causes the body to become dehydrated. This is detected by the osmoreceptors in the hypothalamus which send a response to the pituitary gland to secrete ADH. ADH increases permeability of the walls of the collecting duct which causes more water to be reabsorbed by the kidneys via osmosis, resulting in less urine being produced so less water is lost.

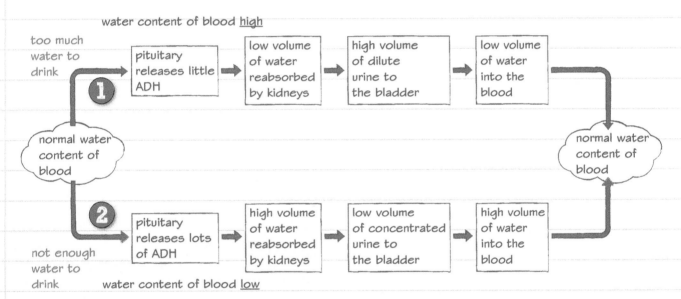

Osmoregulation is important for keeping cells working. If the concentrations of solutes surrounding cells becomes too high (**hypertonic**) the cells will shrivel, or if they become too low (**hypotonic**) the cell will take in too much water and burst.

Now try this

1 Name the gland that produces ADH.

You are working for the RSPCA and have been called out to rescue a dog which has been exposed to very hot temperatures. Upon looking at the dog you find that it is very dehydrated.

2 Explain the effect of dehydration on kidney function.

Nitrogenous waste removal

If there were a **build-up** of **waste** products in the body then these could be **poisonous**, and could **cause damage** to cells, so they need to be removed via **excretion**.

Amino acids and deamination

Some animals' diets can contain a lot of **protein**. This can be used for repair, growth and energy production. However, proteins are made up of amino acids and, when they are being broken down, the **amino group** (NH_2) is removed (deamination) and converted to **ammonia**, a form of **nitrogenous waste**. Ammonia is **highly toxic** and needs a lot of water to **dilute** it for **excretion**, so **land-dwelling** animals convert it into **urea**.

amino group

An amino acid.

Ornithine cycle

The **ornithine cycle** takes place in the **liver**, converting **ammonia** to **urea**. Urea is much less toxic and doesn't need so much water during excretion.

Uric acid

Birds convert nitrogenous waste to insoluble **uric acid**, which is nearly **non-toxic**. Though it uses more energy to make than urea, it requires less water to produce, meaning that birds require less water in their body and are **lighter** for flight.

Liver damage

Cirrhosis is where **scar tissue** affects normal **liver** function. This is a huge problem as it **cannot** be **reversed** and can actually stop the liver from functioning, resulting in **liver failure**. This would affect the ornithine cycle (see above), and cause toxic chemicals to build up.

Common causes of cirrhosis are:

- damage to the liver from disease, medication or toxins
- infection of **hepatitis C**
- excess **fat** build-up in the liver.

There is **no cure** for cirrhosis, however, **preventative treatment** such as **healthy diet**, **exercise** and treatment of underlying conditions can help prevent it from getting worse.

Symptoms can vary depending on the cause, however they will often include:

- loss of appetite
- weight loss
- vomiting and/or diarrhoea
- lack of energy
- increase of urination.

A dog getting plenty of exercise.

Nephritis

Nephritis is an inflammation of the **kidneys** which leads to problems filtering the blood effectively. It can be caused by disease, medication or some disorders. Symptoms include weight loss, protein in urine, increased thirst, increased urination and vomiting.

Now try this

1. What do bird species convert ammonia to?
2. Where does urea get transported to?

The thermoregulatory system

The thermoregulatory system controls an animal's body temperature for both **ectothermic** and **endothermic** species.

Homeostasis

Homeostasis is the process by which the **internal environment** of an organism is controlled. Changes that occur in the environment can have a negative effect on the organism, such as damaging cells, so it is essential for the body to respond to changes to help maintain favourable conditions. Responses to changes can be made through **nerves** or **hormones**. Homeostasis allows the control of different factors such as **blood glucose** levels and internal body **temperature**.

Temperature homeostasis

The body **detects changes** and then responds to correct them to a 'set point' by a mechanism known as **negative feedback**.

Control of body temperature

Species that maintain a fairly **constant body temperature** independent of their environment are known as **endothermic** and this temperature usually ranges from 35–40°C.

Body temperatures of different species.

Species	°C
Human	36.1–37.2
Dog	37.9–39.9
Pig	38.7–39.8
Chicken	40.6–43.0

In the body, the temperature is controlled by the **thermoregulatory system** in the **hypothalamus**. Changes in temperature are detected by **thermoreceptors** in the **brain** (this monitors the core temperature) and the **skin** (this monitors the external temperature).

- When an animal gets too hot, the heat loss centre in the hypothalamus is stimulated.
- When an animal gets too cold, it is the heat gain centre of the hypothalamus which is stimulated.

1 A change in temperature **stimulates thermoreceptors**, which trigger **action potentials** to relay information via sensory neurons to the **hypothalamus**.

2 Here, the hypothalamus (co-ordinator) sends **information** via motor neurons to the **effectors**, in this case, the **skin**.

3 A **response** occurs such as sweating and **vasodilation** of arterioles to allow heat to escape from the skin if the body is too hot.

The response stops when the hypothalamus detects that the temperature has returned to its 'set point'.

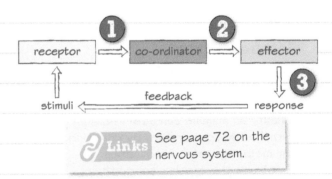

Links See page 72 on the nervous system.

If the **external temperature decreases**, then this causes changes as **arterioles** constrict and hair is erected to trap air. The hypothalamus stimulates production of hormones which lead to an **increase in metabolic rate** and **increased heat production**.

Now try this

1 What is thermoregulation?
2 Name the mechanism that makes changes to allow the body to regulate temperature.

Warming and cooling

Endothermic animals use different warming and cooling mechanisms in response to the environment.

Response to high temperature	Response to low temperature
Vasodilation: Arterioles **dilate**, which allows more blood to enter the capillaries in the skin.	**Vasoconstriction:** Arterioles **constrict**, so less blood flows to the skin.
Pilorelaxation: Pilli erector muscles **relax** which **lowers** the hairs/feathers on the skin so that air can circulate. This allows for convection and evaporation.	**Piloerection:** Pilli erector muscles **contract** causing hairs or feathers on the skin to be **raised**. This traps a layer of insulating air, reducing heat transfer from the skin. **Moulting** can also help.
Sweating: Glands secrete sweat onto the skin, where it evaporates and takes heat from the body. Adrenal and thyroid glands secrete adrenaline and thyroxine which reduces metabolic rate and therefore heat generation.	Sweating, gular fluttering and panting are all **evaporative** cooling mechanisms.
Gular fluttering: Increases heat loss through the mouth. It occurs in birds, where they rapidly flap membranes in the throat to increase evaporation.	**Shivering:** Muscles contract and relax rapidly and cause heat to be produced by friction and respiration.
Panting: Panting is the process of evaporation of water from within the nasal passages, mouth, lungs, and (in birds) air sacs. This method of cooling is used by many mammals, reptiles and birds.	**Brown adipose tissue (BAT):** Cells take lipids and run them through the mitochondria to generate heat. BAT is present in almost all mammals and is found in greater quantities in neonates and hibernating mammals.
Behavioural changes: Stretching out the body gives a larger surface area for heat loss. Animals may seek shade and move less.	**Behavioural changes:** Curling up causes a smaller surface area for heat to escape. Animals may move to warmer areas and show increased movement.

Metabolic rate

Endotherms can adjust their **metabolic rate** to regulate their internal body temperature. Metabolic reactions give off heat, so animals can alter their internal body temperature by changing their metabolic rate. Ectotherms **cannot** use this mechanism on their environment.

Countercurrent mechanism

Some animals have developed a **countercurrent mechanism** to control heat regulation. Heat is exchanged between two sources flowing in opposite directions. In heat regulation, these flowing sources are the bloodstream. When the body temperature drops, arteries carry warm oxygenated blood away from the heart while veins carry cold deoxygenated blood towards the heart. As these two pass, the warm arterial blood transfers most of its heat to the cool venous blood and body temperature is balanced.

Extremes of temperature

If an animal is exposed to extreme external temperature changes, this can result in hypothermia or hyperthermia.

Hypothermia: the body cannot maintain normal body temperature. There are three phases. Symptoms will depend on the phase. **Mild symptoms** include weakness, shivering and lack of alertness. **Moderate symptoms** include muscle stiffness, low blood pressure and slow breathing. **Severe symptoms** include fixed or dilated pupils, inaudible heartbeat, difficulty breathing and coma. Hypothermia can result in death.

Hyperthermia: is the elevation of the body temperature above the normal. **Fever hyperthermia** results from inflammation within the body (e.g. due to infection). **Non-fever hyperthermia** results from all other causes of increased body temperature (e.g. excessive exercise, excessive levels of thyroid hormones). Symptoms include: panting, dehydration, reddening of the gums, decease of urination, uncoordinated movements and unconsciousness. It can lead to multiple organ dysfunction and death.

Now try this

What is the effect of vasoconstriction?

Adaptation and variation

Adaptations to different environments occur over many generations, through **variation**.

Adaptations

Adaptations can be physiological, behavioural and anatomical and can be a result of an animal's lifestyle, diet and/or environment.

- **Physiological** – adaptations in an animal's body processes, for example, some snakes are physiologically adapted to produce venom which they use in defence against predators.
- **Anatomical** – adaptations in an animal's shape and structure, for example, the bat forelimb has become adapted to form wings, improving their access to suitable food sources.
- **Behavioural** – adaptations in an animal's language, tool use and survival strategies, for example, migration in some bird species to avoid harsh environments.

Variations

Variation is the differences between organisms. It can include variations in diet, behaviour and aesthetics. Variation between animals is linked to differences or changes in the DNA. Because of this, variations can be passed on from parents to their offspring, i.e. they are heritable.

Some variations give an animal an advantage over other animals, for example, being able to run for longer, reach higher branches or have better camouflage. Animals with an advantageous variation are more likely to survive and therefore more likely to reproduce. Because the variation is heritable, their offspring are likely to have the advantageous feature, increasing their chance of survival too. This is **natural selection**.

Over time, this can lead to evolution of a species. If animals in a species develop different variations which both give them advantages, it could lead to the species separating into two species; this is called speciation.

Definitions

Evolution – a change in characteristics over several generations.

Speciation – the formation of a new species as part of the evolutionary process.

Polar bear adaptations

☑ Large feet with fur on the soles improves grip on the ice.

☑ Small surface area to volume ratio and small ears reduce heat loss.

☑ Strong legs help catching their prey when swimming and running.

☑ Thick layer of adipose under the skin for insulation and energy.

☑ Thick white fur for insulation and camouflage.

☑ Sharp teeth and claws for feeding and defending.

☑ Pregnant females hibernate.

☑ Highly developed sense of smell and hearing.

Selection pressures on variation

There are different types of selection pressures that affect variation:

1 **Stabilising** selection pressure that reduces variation between animals' DNA. Extreme phenotypes are reduced, and the proportion of animals with the average phenotype is increased.

2 **Directional** selection pressure that encourages a particular new phenotype, usually in response to environmental changes.

3 **Diversifying/disruptive** selection pressure that increases variation between animals' DNA. Extreme phenotypes become more common and average phenotypes are reduced.

Now try this

Red crabs migrate once a year to their breeding ground on the beach.

State the type of adaptation this shows.

Modern technology and classification

Unlike the **binomial system** of naming species using **Latin names**, **cladistics** looks at data from **DNA** and **RNA** sequences, not just physical characteristics.

Cladistics

Cladistics looks at an **evolutionary relationship** between species, not just physical traits. In order for species to be grouped, they must all:

1 share a common ancestor.

2 be included in the same taxon (group).

A **cladogram** emphasises the genetic relationship between species.

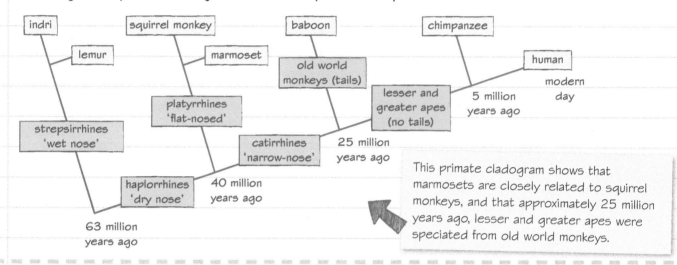

This primate cladogram shows that marmosets are closely related to squirrel monkeys, and that approximately 25 million years ago, lesser and greater apes were speciated from old world monkeys.

Analysing how animals are related

There are different ways to analyse how animals are related.

1 **Genetics:** Use of DNA sequencing to provide information on how species are genetically related and their evolutionary lineage. Comparison of DNA-based sequences can be carried out by:

- **DNA sequencing:** looks at a species' proteins (which are coded for by their DNA). Species that are more closely related will have more similar DNA and proteins than distantly related ones.
- **DNA hybridisation:** looks at how similar DNA is without sequencing it. It involves looking at hydrogen bonds and the amount of heat needed to break them. The higher the temperature, the more closely related the species are.

2 **Biochemical:** looks at the substances found in living organisms and the processes occurring within these organisms.

3 **Immunological comparisons:** analyses antibodies binding to proteins and antigens in different species and if they are recognised. More similar species will have more recognition between their antibodies and antigens.

4 **Amino acid comparisons:** looks at the amino acids formed by the species and the nucleotide sequences they utilise to form the amino acids.

Now try this

1 Explain what information is needed to produce a cladogram.

You are working in a laboratory looking at the genetic relationship between two aquatic species. The process you are asked to use involves the use of heat.

2 State what process you are going to use.

Living organism classification

Animals can be **classified** into **groups** so that they can be **identified**. It also allows scientists to **study** the **relationships** between organisms.

Grouping organisms

Taxonomy is a way of classifying living things based on the similarities and differences between them. There are seven levels or steps of classification, from general to specific. Different characteristics are assessed at each level, and organisms with similar characteristics are grouped together.

Kingdom

Phylum

Class

Order

Family

Genus

Species

> Kingdom is the least specific level of classification, species is the most.

> A **species** is a group of animals with similar characteristics which can interbreed to produce **fertile offspring**.

> This classification system can be abbreviated to: K P C O F G S system.

Kingdom categories

The kingdom level is divided into categories:

Animalia (all multicellular animals)

Plantae (multicellular, all green plants)

Fungi (multicellular, moulds, mushrooms, yeast)

Prokaryotae (usually unicellular with **no** nucleus, e.g. bacteria, blue-green algae)

Protoctista (unicellular with a nucleus, e.g. Amoeba, Paramecium).

> Note the binomial name is written in italics. If handwritten, then underlined.

Binomial names

Every organism is given two **Latin names** (the **Binominal System**):

- genus (upper-case first letter)
- species (lower-case first letter).

For example, *Ursus maritimus* is the binomial name for a polar bear.

Example: Taxonomy of the dog

This example shows the taxonomy of the dog, *Canis lupis*.

A **phylogenetic tree** shows how closely related the organisms are. It shows that some species share a common ancestor: the **ancestor**, and all its **descendants**, are together known as a **clade**.

Kingdom	Animalia
Phylum	Chordata
Class	Mammalia
Order	Carnivora
Family	Canidae
Genus	Canis
Species	lupis

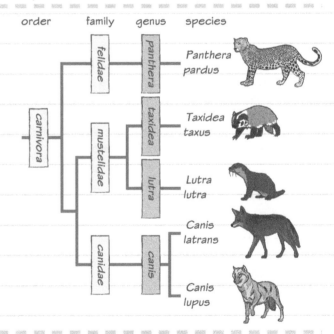

order　family　genus　species

carnivora

felidae — panthera — *Panthera pardus*

mustelidae — taxidea — *Taxidea taxus*

mustelidae — lutra — *Lutra lutra*

canidae — canis — *Canis latrans*

canidae — canis — *Canis lupus*

Characteristics of the five vertebrate classes

1 **Pisces:** gills, wet scales, ectothermic

2 **Reptilia:** dry scales, egg laying, ectothermic

3 **Amphibia:** egg laying, moist skin, ectothermic

4 **Aves:** egg laying, feathers, beaks, endothermic

5 **Mammalia:** live bearing, milk producing, endothermic

Now try this

1 Name **two** kingdoms other than Fungi, Plantae and Protoctista.

2 Write down the **seven** steps of animal classification.

Classification difficulties

Classification difficulties can occur in unusual mammals, for example, the armadillo, bat, duck-billed platypus, whale and pangolin.

Armadillo

Kingdom	Animalia
Phylum	Chordata
Class	Mammalia
Order	Cingulata
Family	Dasypodidae

Armadillos' outer 'armour' is made from specialised bone covered in small overlapping scales made from epidermal tissue. Even though it looks like those within the Reptilia class, an armadillo is actually a mammal.

Bat

Kingdom	Animalia
Phylum	Chordata
Class	Mammalia
Order	Chiroptera

Although bats are within the mammalian class, they have some characteristics of the Aves class. Bats are the only mammals capable of true flight (not gliding). Unlike birds, their wing structure is maintained by the phalanges which move to move the wing. Bats are also split into different family groups.

Duck-billed platypus

Kingdom	Animalia
Phylum	Chordata
Class	Mammalia
Order	Monotremata
Family	Ornithorhynchidae
Genus	Ornithorhynchus
Species	Anatinus

The duck-billed platypus has many characteristics of the Aves class: it has webbed feet and gathers its food underwater. The females lay their eggs in burrows. Around 10 days later, the eggs hatch and the mother will care for them until they are able to swim on their own. There are only five mammal species within the Monotremata order.

Whale

Kingdom	Animalia
Phylum	Chordata
Class	Mammalia
Order	Cetartiodactyla

Whales are mammalia which live in the sea, as opposed to the Actinopterygii class. They are warm blooded and have a four chambered heart, unlike fish which have a two chambered heart.

Other taxonomic strategies

The K,P,C,O,F,G,S system of classification looks at the homologous features between animals to determine common ancestry. An alternative system is to look at analogous features which look similar but does not mean the animals have a similar ancestry.

Pangolin

Kingdom	Animalia
Phylum	Chordata
Class	Mammalia
Order	Pholidota
Family	Manidae

Pangolins are within the mammalian class, although they have similar characteristics to the Ava and Reptilia classes. They are covered in keratin scales and their stomach also has keratinised spines, which project inwards and often contain stones. This helps to grind the food in a similar way to a bird's gizzard.

A pangolin

Now try this

Armadillos have a feature which is not commonly associated with mammals.

Describe this feature and identify which groups of animals this is most often associated with.

Your Unit 2 exam

Your Unit 2 exam will be set by Pearson and could cover any of the essential content in the unit. You can revise the unit content in this Revision Guide. This skills section is designed to **revise skills** that might be needed in your exam.

Short-answer questions
Revise answering these on page 108.

'Complete' and 'define' questions
Find out more on page 109.

Data questions
These are covered on page 110.

Exam skills

'Explain' questions
Revise answering these on page 111.

'Compare' questions
Revise how to tackle these on page 114.

'Discuss' questions
Covered on page 113.

'Describe' questions
See examples on page 112.

Check the Pearson website

The questions and sample response extracts in this section are provided to help you to revise content and skills. Ask your tutor or check the Pearson website for the most up-to-date **Sample Assessment Material** and Mark Scheme to get an indication of the structure of your actual paper and what this requires of you. The details of the actual exam may change so always make sure you are up to date.

Spelling, punctuation and grammar

You should always make sure that your answers are clear, your handwriting is legible, and that you use accurate spelling, punctuation and grammar.

✓ **Spelling** – make sure you spell scientific and technical terms correctly, such as 'integumentary system'.

✓ **Grammar** – in longer answers, make sure you structure your answer clearly. For example, using paragraphs.

✓ **Punctuation** – use commas and full stops to make sure the meaning of your answer is clear.

Worked example

A conservation scientist working at the Game and Wildlife Conservation Trust is surveying a habitat to monitor the water vole population present, as the UK water vole population has been in steady decline.
Scientific taxonomies are used to classify animals to species level.

Define what is meant by the term 'species'.   2 marks

Sample response extract

A group of organisms that are biologically similar and are able to interbreed to produce fertile offspring.

Short-answer questions require you to answer the relevant point clearly and briefly without a long discussion. This answer does this by getting straight to the point.

Your tutor or instructor may already have provided you with a copy of the Sample Assessment Material. You can use this as a 'mock' exam to practise before taking your actual exam.

Now try this

Visit the Pearson website and find the page containing the course materials for BTEC National Animal Management. Look at the latest Unit 2 Sample Assessment Material (SAM) to get an indication of:

- the number of papers you have to take
- whether a paper is in parts
- how much time is allowed and how many marks are allocated
- what types of questions appear on the paper.

Short-answer questions

The command words 'state', 'give' and 'identify' require you to quickly recall facts or main features relating to animal biology. They do not require you to go into a lot of detail, so make sure you get straight to the point!

'State', 'give', 'label' and 'identify' questions

- Questions with these command words are typically worth 1 or 2 marks.
- There is no need to expand to provide more detail.
- Do not spend too much time on your answers.
- If you are not sure, give your best guess and move on to a new question.
- 'Label' questions will usually ask you to provide labels for a diagram.

Worked example

State **two** types of adaption that allow an owl to be an efficient nocturnal predator.   2 marks

Sample response extract

Auditory and olfactory adaptations

This question does not ask you for any detail. In your exam, check the marks available to indicate how many examples you should give.

You only need to name the types of vision for this answer. You do not need to explain what they are.

Worked example

Give the name of the type of vision associated with predator animals and the type of vision associated with prey animals.   2 marks

Sample response extract

(i) Predator animals

Binocular vision

(ii) Prey animals

Monocular vision

Links You can find out more about vision and the structure of the eye on pages 75–77.

Worked example

The diagram shows the internal organs of a cat.

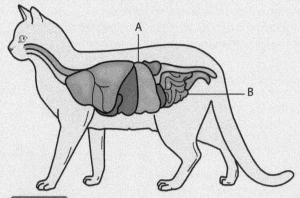

Label the digestive organs A and B.   2 marks

Sample response extract

Organ A
Stomach

Organ B
Intestines

 For 'label' questions, you can give one-word answers.

Now try this

1   Give the names of the four parts of a ruminant's stomach.
2   Identify which part of a ruminant's stomach is known as the 'true stomach'.

Links To revise the digestive system, see page 66.
See page 67 for ruminants.

'Complete' and 'define' questions

'Complete' and 'define' questions are factual questions that require short answers.

Answering the questions

For '**complete**' questions you should:

- provide all the missing items
- look at any examples already in place to see how much you need to write, for example, just one word, a phrase or a sentence.

For '**define**' questions you should:

- clearly state the meaning of the term, for example, by describing what the item is, how it works or what it is like
- make learning the definitions of key terms part of your revision.

Worked example

Organelles are cell structures that are specialised to carry out a particular function or job within a cell. The table below shows some key organelles within an animal cell, with their functions.

Complete the table with the **two** missing organelles and the **two** missing functions. 4 marks

Sample response extract

Organelle	Functions
Nucleus	The largest organelle in a cell. It contains a dense structure called the nucleolus. Contains genetic material, which controls the activities of the cell.
Cytoplasm	Refers to the jelly-like material with organelles in it. Most chemical processes take place here, controlled by enzymes.
Cell membrane	The permeable barrier that controls the movement of substances into and out of the cell.
Mitochondria	These round double membrane-bound organelles are responsible for aerobic respiration. Most energy is released by respiration here.
Ribosomes	These translate genetic information in the form of RNA into proteins. This process is known as protein synthesis.

Links To revise organelles, see page 46.

Worked example

In order of increasing complexity, multicellular organisms consist of:

organelles → cells → tissues → organs → organ systems

Define the word tissue in relation to multicellular organisms. 3 marks

Sample response extract

Most multicellular organisms are made up of different cell types that are specialised to carry out a specific function. These are grouped together and are called tissues.

There are four main types of tissue: muscle, epithelial, connective and nervous.

This answer contains a clear description of what tissues are and what the main types of tissue are.

Links To revise the structure, function and location of tissue types, see pages 52–56.

Make sure you attempt an answer in all the spaces available.

Links To revise the different types of joint, see page 61.

Now try this

The table below gives the names and definitions of different types of joint.

Complete the table with the missing name and the missing definition.

	Bones are held together by fibrous connective tissue which does not allow movement between them.
Synovial	

Data questions

In your exam, you may be required to interpret **data** such as graphs and charts. It is important for you to read the questions carefully and ensure that you have completed all parts.

Representing data

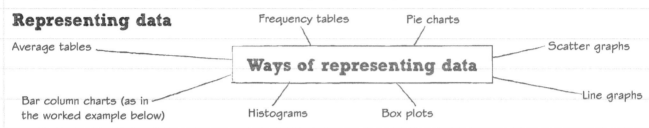

Ways of representing data
- Frequency tables
- Pie charts
- Average tables
- Scatter graphs
- Bar column charts (as in the worked example below)
- Histograms
- Box plots
- Line graphs

Worked example

A dog owner is training her three dogs to enter in agility events. She has taken the dogs' heart rates when resting and exercising. The graph shows the variations in the heart rates.

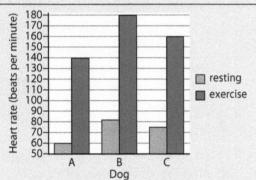

Remember to look at the graph carefully to ensure you have all the information you need.

Calculate the mean average heart rate for these dogs when they are resting and exercising.

(i) Resting   **1 mark**

You need to be able to interpret the graph to find the resting heart rates of the three dogs. Then calculate the **mean** heart rate of the three dogs at rest.

(ii) Exercising   **1 mark**

Sample response extract

Dog A: 60 beats per minute

Dog B: 82 beats per minute

Dog C: 75 beats per minute

$$\frac{60 + 82 + 75}{3} = 72 <72.33333>$$

72 beats per minute

Sample response extract

Dog A: 140 beats per minute

Dog B: 180 beats per minute

Dog C: 160 beats per minute

$$\frac{140 + 180 + 160}{3} = 160$$

160 beats per minute

Take care when calculating your answer. Make sure that you note down the numbers and then go back and double-check.

Always include any **units** in your answer. Check if you need to give your answer to a given number of decimal places (you don't here).

Now try this

You are working at a kennel for working dogs. You are asked to record the heart rates of five dogs before and after exercise in order to monitor their fitness.

Your findings are given in the table.

Calculate the **mean** heart rate **after** exercise to **one decimal place**.

Dog	Beats per minute (bpm)
A	75
B	78
C	82
D	90
E	88

Links To revise changes in heart rate in animals, see page 81.

'Explain' questions

Your Unit 2 assessment is likely to include 'explain' questions. In these, you need to show that you understand the subject as well as giving evidence to support your answer. Make sure that you cover the number of examples given in the question. Three examples are given below.

1 Explain **two** adaptations of polar bears to cold conditions.

2 Explain **one** positive and **one** negative effect for a predator of having binocular vision.

3 With reference to animal adaptations, explain **two** differences between predator and prey animals, giving examples.

A question will usually, but not always, start with the command word. Read the question carefully to find the command word if this is not the case.

The top two questions ask you for two examples. You need to give the examples asked for and give a reason why you have chosen each answer.

Answering 'explain' questions

Make sure you:

☑ Use any data or information that you have been given.

☑ Show that you are reasoning the answer out.

☑ Link observations from the data to your explanation, using words like 'because'.

Worked example

The classification of a guinea pig is:

Animalia, Chordata, Mammalia, Rodentia, Caviidae, Cavia, *Cavia porcellus*.

Explain what classification is and its importance.   **4 marks**

The text above the question will often help to jog your memory about the topic.

Sample response extract

Classification is the process of sorting living organisms into groups. They are classified into groups by physical characteristics and the latest advances in classification allow sorting by molecular biology. Classification is important because it allows the identification and naming of species. It also allows scientists to predict the classification of newly discovered species and trace evolutionary links. The taxonomic classification system is recognised worldwide so research can be shared and it ensures the same species are being studied.

The student gives a clear definition of classification and explains several reasons why it is important.

 Links You can revise taxonomic classification on page 105.

Now try this

Here you need to state at least one alternative theory that could have resulted in the evolution of the new species and give a reason why it may have happened.

A new species of bird has evolved in the rainforest. It was not physically separated from the other birds.

Explain how else this new species of bird could have evolved.

 Links You can revise evolution on page 103.

'Describe' questions

In 'describe' questions, you will be expected to show your knowledge and understanding of the subject.

Examples of 'describe' questions:

1 Describe **two** processes that are involved in digestion.

2 Describe the process of gluconeogenesis.

> If you are asked to describe a specific number of features, you should write one or two sentences for each.

> You may be asked to describe a particular process or topic. If the question doesn't give a number of examples of features to include, you should write between three and six sentences to make sure you cover all the important points.

Answering 'describe' questions

A good answer to a 'describe' question will:

☑ demonstrate accurate and thorough knowledge

☑ apply knowledge to the content of the question

☑ use specialist language consistently and fluently

☑ present the information in a logical order.

Worked example

You are completing work experience at a veterinary practice and are asked to record blood test results and relay the information to customers.

Blood contains three parts: red blood cells, white blood cells and platelets.

Describe the functions of all three components of blood.   `3 marks`

> In this question, you are expected to give a brief description of the function of each component of blood.

> **Links** To revise the composition of blood, see page 79.

Sample response extract

Red blood (erythrocytes) cells deliver oxygen to the body's tissues via a protein called haemoglobin. They also carry back the waste gas carbon dioxide. White blood cells (lymphocytes) defend the body against disease, as they consist of antibodies which initiate immune response. Platelets make up the rest of the blood. They help the body repair by stopping bleeding.

> This student shows understanding by giving a clear description of the functions.

> The student uses the scientific names for the different components, which shows good knowledge.

Now try this

> Here, you are required to state what haemoglobin is and what it does in the body.

Describe the function of haemoglobin.

> **Links** To revise haemoglobin, see page 79.

'Discuss' questions

In a 'discuss' question, the length of your answer will depend on the number of marks the question is worth. You will be expected to write about a subject in detail.

How much to write?

'Discuss' questions may need you to write either a long or short answer. You can use the number of marks for the question to decide how much to write.

- Questions worth around 3–4 marks will need you to discuss straightforward information.
- Questions worth around 8 marks will expect you to discuss more topics in greater depth.

Answering 'discuss' questions

In the answer to a 'discuss' question, you should:

✓ show that you understand the subject

✓ consider all aspects of the question

✓ make connections between different parts of the subject

✓ balance your argument.

Worked example

A zookeeper is asked to brief the team on the impact of environmental changes on animals.

Discuss some adaptations, both physiological and behavioural, that animals have developed to ensure they survive in their environment.   8 marks

Here, you will be expected to explore different aspects of animal adaptation, particularly the importance of animal adaptations for survival.

Sample response extract

Animals that live in very hot climates, such as camels in the desert, have body fat stores so that fat is metabolised and produces water. They don't sweat and produce very little urine, so lose less water. However, some animals need to cool themselves by evaporative cooling, such as birds and animals that have sweat glands.

This student includes an example which shows understanding and also provides a counter example. It is important to look at more than one side of a topic.

Animals that live in colder climates, such as polar bears, tend to have a large body size and a thick layer of fat, which is necessary for insulation. Waterproof, thick fur provides insulation and they are mainly white in colour for camouflage.

The discussion is well balanced and explores a range of adaptations.

Aerial animals have developed excellent eyesight, powerful wings and a streamlined body. Their talons and powerful beaks allow them to catch prey. In contrast, burrowing animals do not have good eyesight and are generally blind. They have a streamlined shape and large front paws for efficient digging. They rely on heightened senses of smell and touch.

This is an excellent comparison of different types of animal.

Links You can revise physiological and behavioural adaptations on page 103.

Here, you need to discuss how being too hot or too cold affects blood vessels and capillaries. Marks will be allocated according to the level of knowledge and understanding that you demonstrate.

Now try this

Discuss the relationship between blood vessels and capillaries in the control of an animal's body temperature.

 Links See page 80 to revise the structure and function of the blood vessels.

'Compare' questions

In 'compare' questions you are asked to point out at least one similarity and one difference. The total number of marks indicates the total number of points you should make.

Answering 'compare' questions

- ✓ Look for the similarities and differences between two or more things.
- ✓ Use a clear structure, e.g. start with the similarities and then move on to the differences.
- ✓ Your answers should be factual and include clear descriptions.
- ✓ You should not have to draw a conclusion.
- ✓ Your answer must relate to all the items mentioned in the question.

Worked example

You have completed a dissection practical and have investigated different types of muscle.

Compare smooth and cardiac muscle.

4 marks

Sample response extract

> The student has provided a similarity, as well as explaining the differences.

Both smooth and cardiac muscle are involuntary muscle – controlled by the autonomic nervous system (ANS). Smooth muscle contains cells that are spindle-shaped with one nucleus, whereas cardiac muscle contains elongated, branching cells with one or two nuclei per cell. Smooth muscle is non-striated, while cardiac contains striations. Cardiac muscle is very specialised and is found only in the heart, while smooth muscle can be found in the digestive system and other parts of the body such as the bladder.

> **Links** To revise muscle types see page 55.

Worked example

Herbivores such as cows eat a plant-based diet, while carnivores such as dogs eat mainly a meat diet.

Compare the dentition of a herbivore and a carnivore.   **4 marks**

Sample response extract

Herbivores include ruminants whose diet mainly comprises of plant material. Carnivores such as lions have a diet that consists of meat.

Both carnivores and herbivores have molars for grinding down their food as the first part of digestion. Both have teeth that are made up of enamel, which is the outer layer of the tooth.

Ruminants find it difficult to digest and swallow food generally if unchewed. Their dentition lacks upper incisors, which are replaced by a 'dental pad'. They use this and their molars to grind regurgitated food and mix it with large amounts of saliva. Carnivores have specific teeth specialised for different tasks. They have sharp canines and carnassial teeth, which can break tendons and bones and shear meat.

> It is a good idea to briefly explain why the dentition is different. Giving a brief definition of herbivore and carnivore sets the context and shows good knowledge.

> If asked to compare, you can highlight some similarities as part of your response ...

> ... as well as the differences in the two types of dentition.

> **Links** To revise dentition, see page 68.

Now try this

Compare the parasympathetic and the sympathetic nervous system.

> **Links** To revise the autonomic nervous system, see page 73.

Animal welfare

All those involved in caring for animals have responsibilities for the **welfare** of the animals. Animal welfare organisations promote and assist people in meeting these responsibilities.

What is animal welfare?

Welfare refers to an animal's health and well-being, and how well it is coping in its living conditions. Many different elements are needed to provide a high level of welfare, including suitable environment, diet and opportunities to socialise. If the correct elements are provided for an animal, it has a good chance of being happy and healthy.

 To revise different definitions of animal welfare, see page 136.

The five animal needs

Under the Animal Welfare Act 2006, owners are legally responsible for making sure the **five welfare needs** of their animals are met:

1 Need for a **suitable environment**

2 Need for a **suitable diet**

3 Need to be able to **exhibit normal behaviour patterns**

4 Need to be **housed** with, or apart from, other animals

5 Need to be **protected from pain, suffering, injury and disease**.

Links To revise how to put the five needs into practice, see pages 116–130.

Animal welfare organisations

These raise awareness of the needs of animals to avoid suffering. They may provide services to help owners meet the needs of their animals. Some organisations, such as the Royal Society for the Prevention of Cruelty to Animals, investigate reported cases of neglect or cruelty, and may bring private prosecutions against people who deliberately cause suffering to animals.

The RSPCA is a well-known UK animal welfare organisation.

Links There is more information about animal welfare organisations on page 135.

Monitoring animal welfare

Welfare appraisals may be conducted by welfare organisations, local authorities or the Animal and Plant Health Agency (APHA). They may be part of a regular programme of inspections, such as for licensing purposes, or in response to reports of animal welfare issues.

Animal welfare inspectors can also carry out random checks. If necessary, they will specify welfare improvements that must be made, such as changing the type of enclosure, and carry out follow-up visits to check these have been done.

All people working with animals should monitor the animals' welfare regularly and respond appropriately to changes in the animals' needs.

Outcomes of animal welfare appraisals

☑ **Educate** owners or businesses on the legal requirements for animal welfare.

☑ **Warn** owners or businesses of the consequences if issues are not addressed.

☑ **Remove any animal(s)** of concern with permission (from owner or local authority).

☑ **Prosecute** offenders where necessary.

Links See pages 151–155 for more information on the welfare appraisal process.

Now try this

Explain **one** method that animal welfare organisations can use to help owners comply with animal welfare legislation.

Animal housing

One of the five animal needs is the **need for a suitable environment**. This includes the type of **accommodation** provided.

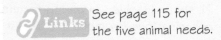 **Links** See page 115 for the five animal needs.

Factors to consider when selecting suitable accommodation

There are several factors to consider when assessing the suitability of accommodation, in order to maintain good levels of animal welfare.

construction materials – suitable for the intended species (e.g. non-toxic)

strong and weatherproof

able to house the animal(s) comfortably

suitable for the intended purpose – e.g. breeding considerations accounted for

species requirements – recommended minimum size specified by the RSPCA

Type of accommodation

Key accommodation considerations

Size of the accommodation

suitable stocking density – e.g. for livestock

Access to the accommodation

Links To revise stocking densities, see page 120.

allows access to water and food according to species' preferences

easy to clean and maintain

doors or lids can be opened and closed without obstruction

Location of the accommodation

predators and prey kept at an acceptable distance from one another

away from busy areas – e.g. traffic, lots of visitors

away from direct sunlight and draughts

Water sources

Natural water sources such as streams or ponds can be visually appealing and promote natural behaviours, but may be problematic for some species and handlers (e.g. animal or handler falling in). Bowls for food and water can easily be tipped over, so selecting a more secure method may be preferred.

Pet owners should research the recommendations for a species before buying their housing.

Examples of accommodation and materials

	Example species	Usual construction materials
Vivarium	• Royal python • Bearded dragon	Glass and/or wood
Aquarium	• Neon tetra • Axolotl	Glass or plastic, depending on size
Inside accommodation	• Rat • Guinea pig	Plastic or metal
Outside accommodation	• Rabbit • Ferret	Metal cages and wooden hutches
Aviary	• African grey parrot • Cockatiel	Usually metal (can be inside or outside accommodation)

Now try this

 'Evaluate' means you need to examine both the advantages **and** the disadvantages.

Evaluate the suitability of the accommodation pictured above as housing for two rabbits.

Suitable environment

In order to have a **suitable environment**, animals need suitable species-appropriate **substrates**, as well as the correct **temperature** and **humidity** for the species.

Substrates

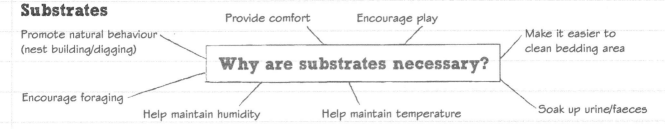

Provide comfort   Encourage play

Promote natural behaviour (nest building/digging)

Make it easier to clean bedding area

Why are substrates necessary?

Encourage foraging

Help maintain humidity   Help maintain temperature

Soak up urine/faeces

Types of substrate suitable for use with different animals

Substrate	Pros	Cons	Suitable for ...
Shavings	• Absorbent • Easily accessible	• Dusty • Can contain splinters	Small and large mammals
Straw	• Very warm • Animals can create nesting areas	• Can appear messy • If stored in sheds can become damp	Large mammals and pets such as rabbits and guinea pigs
Hemp bedding	• Allows burrowing • Dry and dust-free • Absorbent	• Not suitable if damp enclosure required	Snakes, tortoises, small and large mammals
Peat	• Maintains humidity • Easily accessible	• Can be easily ingested	Some amphibian and snake species
Orchid bark	• Maintains humidity well	• Can be ingested • Can be costly	Reptile and invertebrate species
Sphagnum moss	• Absorbs and retains water	• Can harbour bacteria if used for long periods	Reptile and invertebrate species
Shredded paper	• Cheap • Extra warmth	• Can appear messy • Can be ingested	Small mammals
Aquarium gravel	• Aesthetically pleasing • Offers feeding enrichment	• Time consuming to clean thoroughly	Aquatic species: fish and some amphibians

Alternative substrates may be used, such as veterinary bedding in a veterinary surgery.

Temperature and humidity requirements

In the UK, temperature and humidity can vary considerably, which affects how we house animals.

Reptile and amphibian enclosures need heat lamps or mats and a small vent, particularly in winter; a water source or misting to maintain humidity, particularly in summer.

Mammal and bird enclosures are often open with provision for animals to cool down or keep warm as necessary (e.g. substrates, bathing area). In winter, animals may need more substrate or animal blankets (depending on species).

Definitions

Exothermic ('cold blooded') animals (e.g. reptiles): cannot regulate their own body temperature.

Endothermic animals (most mammals and birds): able to generate their own internal heat.

The naked mole rate is the only known **poikilothermic** mammal – unable to regulate its own body temperature.

Now try this

You work in a collection that houses ferrets outside. The temperature at night is 5–8°C.

Discuss the heat sources and substrates you would provide for the hutches housing the ferrets.

117

Balanced diets

One of the five animal needs is the need for a **suitable diet**.

 Links See page 115 for the five animal needs.

What is a balanced diet?

Animals require a **balanced diet** – the correct combination of foods to provide the **nutrients** they need to thrive. This will depend on the **species, life stage**, and types of feed available.

Nutritional value versus palatability

Animals find some foods more **palatable** (more appealing smell, texture and taste) than others, for example, fresh foods or wet foods rather than dry foods. To ensure the **nutritional value** and to prevent deficiencies, avoid topping up mixed feeds until all components are eaten.

Links To revise dietary requirements for specific species, refer to pages 119–121.

Choosing feeds

Animal keepers need to consider feed **types, costs** and **storage** methods.

Feed (whether meat or vegetable based) is commonly sold as dry, fresh, frozen or live, depending on the species being fed.

Frozen feed is easy to store and may be cheaper than fresh feed. Pelleted feed may contain higher nutritional values than similar quantities of fresh or baled feed.

Large sacks of pelleted feed should be stored in airtight containers to prevent animals, flies and damp getting in.

 Links To revise **concentrate** and **forage** **ration** formulation, see page 120.

The seven main nutrients

Nutrients	Example sources	Key functions	
Carbohydrates	Vegetables, legumes, rice, grains, fruits	Energy source	
Protein	Eggs, meat, fish, legumes, seeds	• Growth and repair • Potential energy source	
Water	Vegetables, fruits, water	• Lubrication and hydration • Removal of waste • Allows chemical reactions to take place	
Fibre	Hay, hulls of grains and seeds, nuts	• Aids movement of food through digestive system • Expulsion of waste material	
Fats (lipids)	Meat, fish, nuts, seeds, oils	• Energy source • Energy storage	• Insulation • Cushioning of organs
Vitamins	Fish, vegetables, fruits	Important in chemical reactions in cells	
Minerals	Fruits, vegetables, meat	Many functions in overall body health	

Nutrition at different life stages

An animal requires **different** levels of nutrients as it moves through the **life stages**.

Working animals require large amounts of carbohydrates because they move a lot and use lots of energy.

Lactating animals require more calcium because they are producing calcium-rich milk to promote bone growth and development in their young.

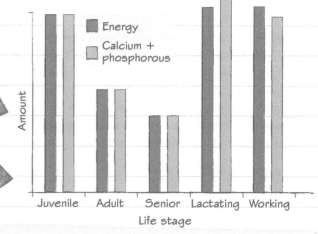

Requirements for energy, calcium and phosphorous.

Now try this

Explain why a balanced diet is important for animals.

Nutritional problems

Consideration of a **balanced diet** is very important in order to avoid nutritional problems in the animals you care for.

Causes of nutritional problems

Nutritional problems such as **diarrhoea**, **diabetes**, **obesity** and **nutritional deficiencies** may occur in **captive** and **domestic animals** that do not have access to natural diets.

Owners should learn about their animals' dietary requirements, and avoid sharing human foods, feeding diets meant for other animals or offering too many treats.

> Many 'human' foods, such as chocolate, onions and xylitol (an artificial sweetener) can be very toxic to animals.

Prevention and treatment of nutritional problems

☑ Provide the **correct nutritional balance**.

☑ Provide **supplements** where necessary.

☑ Make sure **water** is always available.

☑ **Avoid** giving human foods and extra treats.

☑ Provide **sufficient exercise**.

☑ Provide **medication** where necessary, e.g. insulin for diabetic pets.

🔗 **Links** To revise balanced diets, see page 118.

What is supplementation?

Dietary supplements, usually **vitamins** and **minerals**, are added to an animal's food or water to help provide a balanced diet and prevent common nutritional problems. They provide nutrients that may otherwise be missing from the diet, or that the animal cannot consume in sufficient quantities.

Make sure you know which supplements have been scientifically established as essential in the diet, and which are marketed as promoting health, but without clear evidence.

Specific dietary needs

Different species have very specific dietary needs.

- **Taurine** is an essential amino acid found in animal protein and needed by **cats**. Almost all commercial cat food is enriched with taurine. Excessive amounts of vitamin A are toxic for cats.

- Cats that are fed dog food can, over time, develop severe problems with their eyesight, heart function and immunity, because their dietary requirements are so different.

- **Guinea pigs** must be fed diets high in **vitamin C**, because they cannot produce their own.

Metabolic bone disease (MBD)

Reptiles need **Vitamin D_3** to aid absorption and utilisation of the important minerals **calcium** and **phosphorous**. A deficiency in these can cause reptiles to develop brittle (easily broken) or bendy bones.

Prevention of MBD is simple – provide the correct diet and environment, including:

☑ suitable UVB lighting that allows reptiles to produce their own Vitamin D_3

☑ correct supplementation, for example, calcium and phosphorus.

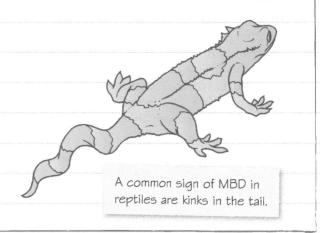

A common sign of MBD in reptiles are kinks in the tail.

Now try this

> A reptile keeper is planning to keep bearded dragons.

1 What nutritional problems should they be aware of?   2 How can they avoid these problems?

Dietary needs of ruminants

Ruminant animals have a different digestive system to other types of animal and therefore have specific **dietary requirements** that must be met.

A ruminant's digestive system

Ruminants are a special type of herbivore with four stomach compartments instead of one. They need access to the **correct foods** for the **correct amount of time**.

Ruminants are foregut fermenters, which means most digestion occurs at the front of the digestive system. They extract nutrients by passing partially broken down food back and forth between the stomach compartments, which is a long process. They need to **graze** for most of the day to obtain full nutrients from their food. Owners must provide sufficient space and food for the number of animals, e.g. field space with good pasture, plus silage and hay.

> **Links** See page 67 for information on the digestive system of ruminant animals.

See page 67

Common ruminant species

These include:
- cattle
- sheep
- goats
- buffalo
- deer
- giraffes.

Meeting the dietary needs of cattle

Cattle should have a good-sized **grazing area** which is safe and secure with good-quality pasture. They need access to a **sheltered area**, and there should be a 'penned' area for observing the animals (to be used if there are concerns over food intake, etc).

They will also require an eating area where **feed** is provided in a trough. If the type of feed needs to be changed, this should be done gradually. Concentrate feeds should be considered for times of drought, or when trying to 'bulk up' animal weights, or during lactation. These can be scattered on pasture or provided separately in the trough.

High-quality pasture means a greater proportion of the nutritional value can be used, for example 80 per cent used by the animal, with just 20 per cent passed as waste.

Definitions

Pasture quality: the amount of nutritional value utilised by the animal. Silage and haylage are pasture which has been harvested in the summer months when it is high in nutritional value, then stored. These are fed to ruminant animals when pasture is not as good, for example, in the winter. The quality of both must be high.

Stocking density: the number of animals grazing on a specific area, for example, in a hectare, at one time.

Wet conditions and many animals trampling an area cause compaction, and can limit the growth of grass and cause weeds to appear.

Concentrate feeds for farm livestock

Concentrate feeds provide concentrated sources of necessary nutrients for livestock, either as complete diets or to supplement **forage rations**. The ingredients will vary for different species and requirements (e.g. lactating animals). They generally contain cereal grains and pulses in different ratios (to provide energy and protein), often with added minerals and vitamins.

Concentrate feeds can be bought ready prepared or can be created on the farm, however, this requires knowledge and skill. Pearson square ration formulations can be used to calculate concentrate feed rations.

> **Now try this**
>
> Define what a ruminant animal is, giving **two** examples of species.

Dietary needs of hindgut fermenters

Hindgut fermenter animals have a different digestive system to other types of animal and therefore have specific **dietary requirements** that must be met.

A hindgut fermenter's digestive system

Hindgut fermenters are herbivorous animals with a special type of digestive system.

They have just **one stomach compartment**, and cellulose is broken down **after** the food has passed through the stomach (in the caecum and colon), rather than before as in ruminant animals. The stomach compartment is often small, particularly in comparison to the stomach of a ruminant animal.

The **caecum and colon are enlarged** in order to store microbes which allow fermentation to take place, and to allow further nutrients to be absorbed.

> 🔗 **Links** See page 67 for more information on the digestive system of hindgut fermenters.
> For ruminant animals, see page 120.

Common hindgut fermenter species

These include:

- rodents (for example, rats and guinea pigs)
- horses
- elephants
- rabbits
- rhinos
- zebra.

Meeting the dietary needs of horses

Horses need good-sized accommodation, comprising an **inside stabled area** and **outside access** which is safe and secure. Easy access to the stabled area is useful when providing muesli-style feed to observe the horse's intake. Good-quality **grass and hay** should form the bulk of a horse's diet. **Salt licks** can be provided in the stabled area, to provide an enrichment activity as well as the necessary electrolytes for normal body function.

Horses can become bored during stabled times and continue to lick or chew salt blocks, so you could use loose salt to limit the intake.

Horses need frequent access to good-quality hay to ensure sufficient fibre intake.

Caecotrophs

Some hindgut fermenter species cannot absorb enough nutrients from their food, so they produce **caecotrophs** – a special kind of faeces which needs to be re-ingested as soon as its expelled from the body. The digestive system continues to break down and absorb the nutrients that have not been absorbed first time round. If the animal does not re-ingest its caecotrophs, it will not absorb sufficient nutrients, for example vitamins, and its health can suffer. Eating faeces is also known as **coprophagy**.

Elephants and giraffes

Elephants spend most of the day looking for and consuming the large amounts of vegetation they need. They have tough trunks for removing plants and shrubs, and large flattened teeth for grinding plants with different textures.

Giraffes have a similar diet. They tend to eat leaves, flowers, fruits and buds located high up on trees, and do not consume much grass. Feeding on items higher up is better for the giraffes, as it also allows them to look out for predators.

Caecotroph animals are usually from the lagomorph family, for example, rabbits.

Now try this

1 Define what a caecotroph is.
2 Why do some hindgut fermenter species produce them?

121

Normal behaviour patterns

One of the five animal needs is the need to be able to **exhibit normal behaviour patterns**. Each species has specific behaviours associated with different situations.

🔗 **Links** See page 115 for the five animal needs.

What is 'normal behaviour'?

Normal or natural behaviours are those we would expect to see displayed by animals living under normal circumstances in the wild. In captivity, animals must be provided with opportunities to display natural behaviours, although this can sometimes be difficult because of the impact of domestication.

Example: Dogs evolved to live in hunting packs, where competition, aggressiveness and killing prey are natural behaviours, although most pet owners would consider them as unacceptable behaviours.

Abnormal behaviours in captive or domestic animals are likely to be a sign that the living conditions are inappropriate for the species.

🔗 **Links** See page 153 to revise using behaviour to monitor animal welfare.

Examples of normal behaviour

- ✓ Cows **grazing** together.
- ✓ Meerkats **digging** for food.
- ✓ Bearded dragons **basking** in the sun or under a heat lamp.
- ✓ Dogs or cats involved in **grooming**.
- ✓ Horses or monkeys involved in **allogrooming** (the act of grooming another animal).
- ✓ Lions **hiding** in thick vegetation while stalking prey.

Defence behaviours

Animals use **defence behaviours** to avoid being injured or killed by another animal.

Common behaviours include:

- **hiding** from threats, e.g. humans, other animals or unsuitable environments (e.g. noisy places)
- **warnings**, to frighten the predator away, or to allow the prey animal to escape

Species	Defensive behaviour
Porcupine	Raising its spines
Rattlesnake	Rattling its tail
Cat	Hissing, showing teeth
Snakes and lizards	Playing dead to avoid being captured

Courtship and mating

Animals may become more aggressive during **courtship and mating**. Although concerning for animal carers, this is quite normal. For example, male ferrets 'drag' the female around by the scruff of the neck and cause visible bite marks.

Conflict occurs between male animals of many species during this time in order to win the right to mate with the female(s).

Captive breeding schemes

Breeding animals in captivity, such as in zoos and on farms, is usually carefully planned. For example, rams selected for breeding are kept separate from ewes, except for at the correct time of year for mating. This reduces conflict between males and ensures success in breeding.

Predator avoidance

Predator avoidance is a survival technique adopted by many species, for example, alerting other animals in the group by making alarm calls.

Meerkats watch for predator species and alert the group if they see a threat.

Now try this

Choose an animal species and make a list of the normal behaviours for that animal.

Behaviour and feeding

Two animal needs that are very closely linked are the **need for a suitable diet** and the **need to be able to exhibit normal behaviour patterns**. These can be met by providing the correct diet in a way that encourages normal behaviours, where possible.

 Links See page 115 for the five animal needs.

Feeding enrichment

Feeding enrichment means providing feed to a domestic or captive animal in different ways to help stimulate their mind, provide correct nutrition and encourage natural behaviour – all of which will aid in animal welfare.

Links For more on feeding enrichment, see page 125.

Diet types
Some of the main types of animal diet are:
- ✓ herbivore (plants)
- ✓ carnivore (meat)
- ✓ omnivore (plants and meat)
- ✓ insectivore (insects)
- ✓ piscivore (fish).

Normal feeding behaviours

An animal's diet will impact on its feeding behaviours. An example species is given for each behaviour.

Stalking prey
Observing and monitoring the prey animal during hunting (e.g. lions).

Prey capture
Capturing the prey animal during hunting (e.g. wolves).

Striking
Making a sudden, violent action towards a prey animal, to stun or kill (snake species, e.g. Texas rat snake).

Foraging
Searching across an area for food (e.g. baboons).

Selective grazing
Selecting particular plants when feeding on grass or pasture (e.g. goats).

Different feeding behaviours

Burrowing
Burrowing into their food, e.g. eating and tunnelling (e.g. moles).

Rooting
Using the snout to burrow or dig in the ground in search of food (e.g. pigs).

Browsing
Feeding on leaves, fruits and shoots that are high up (e.g. deer).

Filter feeding
Filtering small organisms through an (aquatic) animal's system (e.g. flamingos).

Scavenging
Consuming already dead plant or animal material (e.g. hyenas).

Links Coprophagy is another feeding behaviour, covered on page 121.

Hunting behaviours

It can be difficult to encourage natural hunting behaviours in captivity. Live invertebrate species can be supplied as food items in the UK, but not vertebrates. Animals that naturally hunt vertebrate species must be provided with an **alternative enrichment technique**. For example, captive lions can be provided with meat-filled structures, or meat can be hung high up.

Live invertebrates such as crickets and mealworms can be left in enclosures but should be monitored and not left for too long, in order not to compromise the health of the animals.

Domestic cats and dogs often exhibit hunting behaviours, but usually do not consume any animals they catch.

Food such as humanely pre-killed rodents for snakes can be warmed up and made to look alive by moving them around with tongs.

Now try this

Describe some of the different feeding behaviours that herbivorous animals may demonstrate.

Circadian rhythms

Part of meeting an animal's need to be able to **exhibit normal behaviour patterns** involves enabling it to follow its natural sleep/wake cycles.

What are circadian rhythms?

The changes in an animal's activity levels during the day and night are known as its **circadian rhythm**, commonly known as the **sleep/wake cycle**.

Each species will have certain times of the day and night when they would normally be active and others when they should be asleep.

The three main types of circadian rhythm

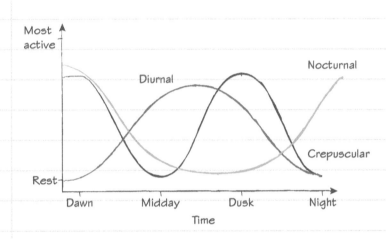

Nocturnal: active through the night and rest during the day, e.g. owls and lemurs

Crepuscular: active at dawn and dusk, e.g. cats and deer

Diurnal: active through the day and rest during the night, e.g. deer and squirrels

Influences on an animal's activity levels

Factors which may disturb an animal's natural circadian rhythm in captivity or a domestic environment include:

1 Environment:
- Temperature
- Enclosure size
- Other animals in enclosure
- Noise
- Light (e.g. too much during the night or too little during the day).

2 Stereotypic behaviour (repeated behaviour with no purpose) caused by environment or other animals.

3 Owner's lifestyle or activity

4 Illness

5 Companionship or lack of companionship (of other animals or humans).

Examples of providing the right conditions

✓ Provide cat flaps so cats can come and go at any time of day or night.

✓ Place hamsters in a room that is quiet during the day so they can sleep.

✓ Dairy cows should only be moved and milked during daylight hours.

✓ Owners who are active encourage more activity from pets, e.g. frequent walks or playing with balls or rope toys.

✓ Seek veterinary treatment for illness promptly and administer medicines as instructed.

✓ If limited human contact is given, social animals must have the companionship of other animals, e.g. rabbits housed in pairs.

Now try this

Define the term nocturnal, giving an example of a nocturnal species.

Environmental enrichment

A range of methods can be used to provide a stimulating environment for domestic or captive animals in order to prevent boredom and encourage **normal behaviour patterns**.

Benefits of environmental enrichment

Environmental enrichment improves the welfare of animals on farms, in labs and in zoos, as well as pets. The main benefits include:

- a more natural way of feeding
- keeps the animal occupied
- mental stimulation
- creates a natural environment
- sensory experiences – sights, smells, tastes
- encourages animals to exercise
- encourages natural behaviours
- encourages social behaviours
- reduces the likelihood of stereotypic (abnormal) behaviour.

Links Animals that do not have enough stimulation can develop stress behaviours – see page 153.

Enrichment structures and devices

✓ **Structures** might include trees or boulders in an enclosure. For example, tree-like structures above head height of humans/potential predators for marmosets to demonstrate normal interactive locomotor play.

✓ **Devices** are items that animals can move, manipulate and play with, such as ropes, tyres, brushes, and purpose-made toys and puzzles.

Budgerigars are intelligent birds. They need enrichment devices to fully utilise their beak and claws, and keep their minds stimulated, especially when housed singly.

Types of feeding enrichment

Ways of enriching the feeding experiences of domestic and captive animals to simulate how they might search for/hunt and eat their food in the wild include:

✓ **freezing** fruits or meat in ice blocks, e.g. for monkey species, polar bears and fish

✓ **location of feed provision:**

  ○ high up in the trees, e.g. for leopards
  ○ in the water, e.g. for otters
  ○ all round an enclosure to encourage flying and moving, e.g. fruit kebabs for parrots

✓ **scatter feeding** and **hiding** food – encourages foraging, e.g. for meerkats and rodents

✓ providing **whole foods** – carcasses for carnivores, e.g. whole turkeys for lions

✓ **replicating the chase** – feeding from the back of a landrover in wildlife parks, e.g. for lions

✓ **treat balls** – toys which reward pets with treats that fall out, e.g. for dogs, rabbits, ferrets

Food toys and puzzle feeders are a great way of enhancing feeding for primates, and many other animals.

Always make sure that devices are safe for the animals to use.

Now try this

1. Explain the benefits of environmental enrichment.
2. Describe methods of feeding enrichment for **two** different domestic animals.

Exercise

One of the five animals needs is the **need to be protected from pain, suffering, injury and disease**. The normal behavior patterns of most animals include movement and exercise.

🔗 **Links** See page 115 for the five animal needs.

Why is exercise important for animals?

- Maintaining its normal exercise levels will help to keep an animal **healthy**.

- Healthy animals will have a **longer life span** than those that do not exercise or have a healthy diet.

- Exercise has both **mental** and **physical benefits**.

- Animals that are able to exercise have **fewer behaviour and health issues** because their minds are occupied, they become tired, and fat stores are not left to build up.

- It also helps **reduce** the chances of animals acquiring **diseases** and developing health conditions such as obesity and diabetes.

Types of exercise provision

Enrichment activities, e.g. puzzles and toys for all kinds of animals

Enclosures that allow space for exercise, depending on size and exercise needs of the species

Structures to encourage climbing, e.g. monkeys

Exercise opportunities include

Walking, e.g. dogs

A pool for swimming, e.g. terrapins

Access to outside, e.g. poultry

Play wheel, e.g. degu

A run, e.g. rabbits

Riding, e.g. horses

Small rodents can be encouraged to exercise by placing a ball or a wheel in their enclosure.

When is exercise not recommended?

Too much exercise can cause problems. Animals that are exercising too hard, for too long or during phases of rapid growth with limited rest breaks may develop problems with joints and feet.

Animals that are ill or injured may need to have limited or even no exercise until they recover:

- **Navicular disease** in horses is a condition of the front hooves that can be worsened by hard exercise, causing the horse to become lame.

- Dogs with **hip dysplasia** (a severe joint condition) will need controlled, low impact exercise such as short walks on the lead and swimming, following an exercise programme provided by a vet.

🔗 **Links** See page 65 for more on hip dysplasia.

Exercise at different life stages

✓ **Senior** and **young animals** generally require less exercise than adults. Younger animals are more prone to developmental problems if over-exercised during growth spurts.

✓ **Working animals** may exercise more (expend more energy) than other animals, and so require more energy from their diet.

✓ **Pregnant and lactating animals** should have reduced exercise as they will need to use their energy for their offspring.

Now try this

Explain the benefits of exercise for animals.

Solitary animals

One of the five animal needs is the need to be **housed with, or apart from, other animals**. Animals' housing needs will differ according to the sociality category they belong to.

Links See page 115 for the five animal needs.

Sociality categories

Animal species can be categorised according to **sociality** (their tendency to form social groups).

Sociality	Presocial	Solitary	Eusocial
Definition	Animals that live together in groups in order to benefit each other	Animals that usually live alone, but come together for mating	Animals that live in colonies and have specific roles
Examples	Badgers, sheep, dolphins, chimpanzees, cockatoos, tetra fish	Tigers, tarantulas, some hamsters, Siamese fighting fish, most reptiles	Hymenoptera (ants, bees, termites, wasps), naked mole rats

Solitary animals

Solitary animals spend most of their lives without others of their own species. They tend to live alone and find their own food. Solitary animals may come together in order to **mate** or when caring for **offspring**. Once the offspring are old enough, or the adult is ready to mate again, the young will be rejected and they will need to fend for themselves.

Tigers are a solitary species, preferring to live alone.

Advantages and implications of solitary behaviour

👍 **Little or no competition** for food and mates.

👍 **No need to share** food once caught/gathered.

👍 Individual animals often have established boundaries or territories in order to **avoid conflict**.

👎 **Fights over territory can occur**, and usually animals will fight to the death or until serious injury occurs.

👎 Choice of mate and availability of food may be **more limited** than for more social species.

Housing for solitary species

There are welfare implications involved in housing solitary animals with other animals for long periods of time. This may occur in animal collections, and in locations such as in pet stores where space may be limited (e.g. when housing hamsters). Provision for **separate housing** must be made for solitary species wherever possible, to ensure they are not put in stressful or potentially harmful situations.

Genetic diversity is a consideration for those wanting to breed solitary animals. Special effort needs to be made to broaden the gene pool as much as possible in collections.

Links See pages 128–129 for more on presocial and eusocial animals.

A Syrian hamster.

Now try this

Describe how you would provide the appropriate social grouping in accommodation for Syrian hamsters.

Presocial animals

Acceptable living conditions are necessary to ensure the welfare of animals. **Presocial** species (also known as **prosocial**) naturally live together in groups, so it is important to house them with other animals to allow the interaction they need.

Considerations when housing presocial species

Those who care for presocial animals need to think about which animals should be mixed with other animals. Usually groups will be of the same species (**intraspecific** groups), but animals of different species (**interspecific** groups) can sometimes live together happily, for example, dogs and cats, zebras and giraffes. It is important to ensure appropriate size of group, and ratio of males to females, to allow the animals to demonstrate their natural social behaviours.

Animals must be monitored to ensure that they are living together harmoniously for the most part, regardless of species or social needs.

Gorillas, like many other primate species, live in social groups.

Benefits of housing animals in appropriate groups

These can include:

- ✓ good mental health – social animals need interaction
- ✓ grooming and bonding
- ✓ animal learning – social animals teach each other important skills, for example, feeding, hunting, foraging.

Provision for grooming

Grooming can be an important part of the natural behaviours of presocial species. Housing may need to allow access to special areas in order to groom (for example, man-made water holes).

Some common forms of social groupings

Animals form groupings for different reasons, all of which are based around the need for **survival**. There are many theories about why each grouping occurs – the table below gives some examples.

	Types of grouping		
	Aggregation	Mating	Colonies
What does it mean?	Groups (e.g. herds, flocks) come together, but don't intentionally interact	Forming a group of the same species for **mating** purposes	Groups form and **interact** for group survival, e.g. detecting predators and warning the group
Theories for why it occurs in the wild	Often because of attraction to food or water resources	Living with others from the same species makes it much easier to find a mate and reproduce	A larger group size means each individual is less likely to be attacked
Example species	Baboons, sparrows	Pandas, sloths, cats	Meerkats, African penguins

In the wild, becoming separated from the group can mean the death of an individual.

It can be very stressful for captive animals who would normally **aggregate** to be kept alone or in small numbers.

Now try this

A safari park wants to add an elephant to its animal collection.

What should the staff consider to allow the elephant to express its natural social behaviours?

Presocial and eusocial animals

Social behaviour can be split into two further categories – **eusociality** and **presociality** – depending on the extent to which a species performs the key social behaviours.

Eusociality

Eusocial animals live in **colonies** and have complex and rigid social structures. They include several invertebrate species, such as hymenoptera (ants, bees, wasps), termites and aphids.

The complex social structure includes several generations of animals. The colony will have a **queen** which is the only reproducing female. Other members of the colony have **specific roles**, such as:

- cooperative brood care
- gathering food supplies
- protecting the colony and nest.

The naked mole rat is an eusocial species. This is very unusual among mammals.

Presociality

Presociality is very similar to eusociality. It is a form of social behaviour that involves aspects of **social hierarchies and working together** – but not in such a complex way as those of eusocial animals.

Example of presociality

Wolves live together in a social group. They **work together** when hunting and looking after the young, they also have a **hierarchy** – wolves that do not fit into this may be forced out of the group or leave to find another group.

Presociality is much more common in higher animals than eusociality.

Differences between eusocial and presocial animals

There are **three** key social behaviours of social animals living in groups:

1. **communal living** in colonies

2. taking **care of others' young**

3. **sharing the labour** (e.g. protecting others in the group and gathering food).

All three behaviours are exhibited by eusocial species. Presocial species show **some** of the behaviours but not all three. For example, animals in a presocial group may not have clearly defined roles and they may leave to join other groups – this does not occur with eusociality.

Some species of presocial animals live in **colonies**, but they do not demonstrate all the behaviours seen in the members of eusocial colonies.

Housing social animals

When planning to meet the needs of social animals (whether they are a pet, in a zoo or on a farm), a major consideration must be ensuring that they have **sufficient access to their specific social grouping**.

Housing should always be big enough to accommodate a size of group or colony that allows animals to interact normally.

Now try this

Explain how the needs of presocial animals are different from those of eusocial animals.

 Make sure your answer focuses on the three criteria for social behaviour.

Preventative care and treatment

One of the five animal needs is the **need to be protected from pain, suffering, injury and disease**. In order to keep animals healthy, it is often necessary to use preventative care and treatments.

🔗 **Links** See page 115 for the five animal needs.

Vaccinations

Vaccinations offer immunity to specific diseases and are a form of **preventative care**. Vaccinations should be administered by a trained person.

There is usually a series of injections given over a length of time, with boosters often required to ensure continued immunity.

Example: vaccines for cats

The annual 'booster jab' given to cats ensures they are protected from fatal diseases. Core vaccines include:

- FPV – Feline Panleucopaenia Virus
- FCV – Feline Calicivirus
- FHV – Feline Herpes Virus.

Other vaccinations include Feline Leukaemia Virus (FeLV).

Regular check-ups

Pet animals should have a **veterinary check-up** at least once a year to identify any health problems, usually scheduled to coincide with the animal's annual vaccination. Vets can also offer advice on living conditions, dietary and exercise needs.

Animals held in collections such as zoos are often cared for by an onsite vet.

On farms, the stockperson should check the animals every day. Vets will usually only attend for particular reasons, such as major calving difficulties or diagnosis of disease.

What are parasites?

There are two types of common **parasite** that can affect domestic and captive animals:

- **Ectoparasites** – live outside the host animal, for example, fleas and ticks
- **Endoparasites** – live inside the host animal, for example, tapeworm.

Parasites like ticks are common and treatable.

Parasite control and treatment

Methods of parasite prevention, control and **treatment** include:

- **Insecticide** that kills adult parasites and/or prevents development of their young – for example, collars, sprays, spot-on products, shampoos, tablets or oral drenches.
- **Tick removal** – tools are available to ensure these are removed successfully.
- **Maintenance of the animal's living environment** – cleaning of accommodation, food and water bowls, managing and rotating pasture, etc.
- **Maintaining general good health** – for example, nutrition, foot/hoof care, dental care. Animals with a lowered immune system are more likely to suffer from secondary infections and parasite burden.

Treatment must be carried out regularly, and sometimes more than one method is selected.

Now try this

1 Define the terms endoparasite and ectoparasite.
2 Name an example of each and how it can be prevented, controlled and treated in:
   (i) cattle   (ii) sheep   (iii) rabbits.

Thinking about the meanings of the prefixes 'endo' and 'ecto' or 'exo' in other words may help you, e.g. an exothemic animal requires an external heat source.

What are ethics?

Ethics are a system of moral principles or rules that form the basis of how people make decisions. Ethics are important in animal welfare, as people working with animals often need to make difficult decisions, taking into consideration a wide range of views and opinions.

Factors influencing ethics

There is no one standard set of ethics, as ethics are influenced by many factors.

SOCIAL PRACTICES   PERSONAL FEELINGS

LEGISLATION → **ETHICS** ← SCIENCE

RELIGION

Society and animal welfare

Different countries and societies have different **social practices** (what is considered the 'norm'), including what levels of animal welfare are considered acceptable. Social practices are influenced by:

- whether animals are seen as food sources
- whether animals are used for work or kept as pets
- the level of education regarding animal welfare
- how far the society accepts that animals are **sentient** (have consciousness).

Personal feelings

An individual's **personal feelings** may differ from the social practices of society as a whole. Personal feelings about animal welfare will range from vegetarianism/veganism to not considering animal welfare at all.

Acting **ethically** may sometimes involve putting aside personal feelings for the benefit of animals, for example, euthanising a much-loved but seriously ill pet.

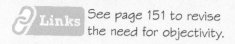

Links See page 151 to revise the need for objectivity.

Science

Developments in **science** influence the ethics of society and individuals, as well as legislation. For example, scientific research into pain perception in animals has shown that vertebrate animals do feel pain in a similar way to humans, whereas invertebrates, such as insects, do not experience pain in the same way. This knowledge influences people's ethical responses to questions such as whether live animals (for example, vertebrates such as mice or invertebrates such as crickets) should be provided as food for other animals.

Links To revise the ethics and legislation covering the welfare of animals used in science, see page 147.

Religious beliefs

Many **religions** consider certain animals to be sacred (e.g. Hinduism, Buddhism), or teach that humans should not inflict harm on animals (e.g. Islam). For example, Muslims believe that Allah created all animals and they should therefore be treated with kindness and compassion.

Some religions teach that animals are equal to humans and they should not be put through pain or suffering, whereas others place humans at the top of the hierarchy. Some religions have changed their teaching over time, and now take a much more animal-friendly approach.

Animal-related legislation in the UK

A country's animal welfare **legislation** (laws) will reflect the generally held ethical views in that country. But legislation is also separate from ethics. Whatever a person's ethical beliefs or personal feelings are, they must abide by the law of the land. The UK animal welfare laws were introduced to establish welfare standards for animals in settings such as zoos, farms, pet stores, laboratories and domestic pets.

Links See page 137 and pages 140–150 for animal-related legislation.

Now try this

1. Explain how personal feelings and social practices can be different, with an example.
2. Explain why animal welfare legislation exists. Which animals are covered by legislation?

Ethical theories

Ethical theories are different **approaches** to deciding what is right and wrong. They can be useful to help us consider different points of view when making decisions about difficult issues.

An overview of ethical theories

Respect for nature – focuses on the importance of species or groups of animals rather than individuals.
- Ethical to contribute to the success of a group of animals, e.g. protect a species from extinction.
- Unethical to genetically alter animals for their survival (because of the need to respect the natural order).

Utilitarianism – considers the sentience of the people and animals involved. Ethically right decisions result in the most happiness for the most people: actions contribute towards the greater good.

Contractarianism – relative cost and benefit is the only thing that matters in the decision-making process. The consequences for the animals do not play a part in the decision-making.

Consequentialism – theories based on decisions about the consequences (outcome) of that decision. The right end result justifies the actions.

Ethical theories

Deontology – also called **duty ethics**; theories are based on the idea that we have a moral duty to behave in a particular way, focusing on the actions rather than the consequences. **The right end result is never justification for actions which are 'wrong'.**

Animal rights – there are strict limits on what we should do to or with animals, no matter how beneficial the outcome may be to humans, because they are just as important as humans.

Environmental ethics – concerns the relationships between humans, animals and other elements of the environment (e.g. trees, mountains, oceans). Considers the question: **Do humans have a moral duty to protect the environment and animals?**

Relational importance – we feel more strongly about animals we have a relationship with (e.g. with pets, or a stockperson with cows). Not many people have a relationship with 'vermin', so don't mind if they are killed as pests.

Sentience-centred – deciding how to treat animals is based on how sentient they are.
Note: Deciding how sentient an animal is may be controversial. It considers whether animals can experience pleasure and pain or feel and think.

Human-centred (also called **anthropocentrism**) – humans are the most important species on the planet and have no duty to care for other species or to preserve nature.

Environment-centred – all living things are dependent on one another and humans are not 'worth' more than other species. Humans must respect all other elements of nature for life to thrive.

The diagram shows some of the ways that people think about the world. There are many more moral positions that have been argued.

 Links What is ultimately accepted as being 'right' is influenced by the factors discussed on page 131.

Choose three of the theories shown in the diagram and apply them to the case of animal testing.

Now try this

Consider how animal testing might be regarded according to different ethical theories.

Exploration of ethical frameworks

You need to be aware of how **ethical frameworks**, and constructing an **ethical matrix**, can help to make decisions about difficult **ethical issues**.

Factors influencing ethical frameworks

An **ethical framework** can be used to consider different ethical theories regarding an issue. In practice, most decisions are likely to be influenced by elements from all the theories.

Theory	Consequentialist	Deontologist	Environmentalist
Questions guiding the framework	Will there be a good result which justifies the process?	Is this the right way? (If so, it is your duty to act in this way.)	How does this affect the relationship between humans, animals and the environment?
Example: Should we use 'battery farmed' hens to produce eggs?	• Eggs are produced relatively cheaply and efficiently. • More consumers can afford basic foods. • Hens bred for battery farming are not endangered.	• The hens are not companion animals, so we do not need to consider their living conditions. • Hens should not be farmed at all – they are as important as humans.	• Hens are part of the natural world and our lives are interlinked. • Hens are sentient beings and should be treated accordingly. • Humans should be allowed to obtain eggs however they choose; no duty to look after other species.

Links See page 132 to revise some important ethical theories.

Ethical matrices

You need to be aware of how ethical frameworks, and constructing an **ethical matrix**, can help to make decisions about difficult ethical issues. Creating an ethical matrix encourages people to take into account the interests of all the **stakeholders** (those involved in or affected by the issue) according to different ethical principles, before coming to a reasoned decision.

Respect for ...	Health and welfare	Freedom and choice	Fairness
Consumers	Quality produce and food safety	Options based on information supplied	Food availability Food choice
Farmers	Good income and working conditions	Freedom of action	Compliance with legislation
Farm animals	Animal welfare and conservation	Behavioural freedom	Intrinsic value of the animals

The ethical principles.

This column lists stakeholders – these could be split into further groups, e.g. 'consumers' could be split into different religious or social groups.

This column shows the criteria needed in order to respect each principle for each stakeholder – you would expand on these depending on the particular question asked.

Now try this

Use the ethical matrix to discuss the ethical issues involved in breeding pigs for the food industry.

Ensure you consider all aspects of the ethical matrix when planning your answer, including all the stakeholders: consumers, farmers and pigs.

Using ethical frameworks

Ethical frameworks are like using scaffolding to approach ethical issues. They can help you to make decisions about what to do, or help you decide on your opinion about an ethical issue.

What are ethical frameworks?

An ethical framework allows you to consider what is right and wrong in a particular situation. It takes into account:

- who the **stakeholders** are, and how they are affected by the issue
- the **viewpoints** of different ethical theories.

It can help you come to a reasoned conclusion about what to do, or what your opinion is.

This is the approach used by **ethics committees** when drawing up **guidelines** on animal use – in science, farming and other situations.

Why are ethical frameworks needed?

Ethical frameworks are needed to:

- avoid making decisions that are based on the interests of the most influential or powerful stakeholders
- take into account the welfare needs of animals (who cannot speak for themselves)
- avoid making decisions based on factors such as accepted social practices (the 'norm') or personal feelings or opinions.

⌘ **Links** To revise the factors influencing ethics, see page 131.

Case study: euthanasia

A veterinary nurse has been asked to advise an owner on the possible steps forward for a senior cat suffering from lymphoma, an aggressive cancer. The owner is elderly and has cared for the cat since it was a kitten.

The treatment may be long, painful and there is a 25%–50% chance the cat will not be cured. Without the treatment, the cat will become progressively more ill and in more pain than before.

The veterinary nurse can also offer the option of euthanising the cat.

The cost of treating the cancer is much more than that of euthanasia.

⌘ **Links** To learn more about DEFRA see page 138.

Using a framework to explore ethical issues

If asked the question 'What is the best course of action in the euthanasia case study?' You would need to structure your answer logically and use the correct key terms throughout. You could use the following framework of questions:

- ☑ What are the **possible decisions** that need to be made?
- ☑ **Who** is affected by those decisions (i.e. who are the **stakeholders**?)
- ☑ **How** are they affected?
- ☑ What is the **'best' outcome**?
- ☑ What may happen as an **unintended consequence** of making that decision?

Make sure you include reasoning based on different viewpoints (ethical theories).

⌘ **Links** To revise ethical theories, see page 132.

DEFRA guidelines

DEFRA publish recommendations for the welfare of a variety of species. The DEFRA recommendations for animal welfare include husbandry and health aspects such as foot care.

The health and welfare guidelines for pigs focus on:

- vaccinations, foot care and other routine procedures
- size and features of accommodation and flooring
- mating information for breeding pigs
- feeding, for example, trough size required per pig.

There is no correct answer here, but make sure you take into account the impact on all of the stakeholders (owner, cat and nurse) and give a reason for your answer.

Now try this

Read the euthanasia case study above and advice on using an ethical framework.

What would you advise the veterinary nurse to do?

Consumer ethics

Consumer ethics considers how people's personal ethics impact on the choices they make when buying products and donating money to charities.

Ethical consumerism

Ethical consumerism means trying to maximise the beneficial effects and minimise the harmful effects of our purchases (on people, the environment and/or animal welfare). It may be influenced by personal feelings, for example, buying free range eggs from farms with high animal welfare standards.

 Links Remember that 'ethical' means different things to different people. See pages 131–132.

Anthropomorphism

Anthropomorphism means assigning human motivations and characteristics to animals and interpreting their behaviour through that filter. For example, while humans might enjoy performing in films or driving cars, such activities would be of no value to wild animals, who are more concerned with moving about freely in their natural habitat. Or, humans might **project** their own fears onto animals, rather than seeing the situation from the animals' point of view, when making decisions about sterilising or euthanising their animal companions.

Speciesism

Speciesism means treating certain species (especially humans and 'higher animals' such as dogs and horses) as more important than others. For example:

If both an animal and a human need to be rescued, should the human be rescued before the animal?

When should humans intervene to help conserve species, and which species should get priority?

- **For** speciesism – humans deserve more rights because they are more intelligent and self-aware.

- **Against** speciesism – intelligence is not a basis for assigning rights; all animals should have rights.

Activism and extremism

Both activists and extremists campaign to bring about **social or political change**. They may share ideas that differ from those of the majority.

Activists use vigorous campaigning techniques (e.g. demonstrating, lobbying, awareness raising). They generally take part in peaceful protests.

Extremists hold beliefs that many people consider unreasonable and/or go to extremes, e.g. may hold louder, more aggressive protests, may be prepared to break the law.

Example: An animal rights activist may hold up signs and chant outside an animal testing facility.

An extremist may defy legislation and break into the facility to release the animals in it.

Animal welfare organisations

- People for the Ethical Treatment of Animals (PETA)

- World Wide Fund for Nature (WWF)

- Animal Aid

- Royal Society for the Prevention of Cruelty to Animals (RSPCA) and the SSPCA in Scotland

- Farming and Wildlife Advisory Group (FWAG)

- Compassion in World Farming (CIWF).

COMPASSION in world farming

Now try this

A person playing with a dog interprets the dog showing its teeth as 'smiling'.

Discuss how anthropomorphism is influencing the person's response to the dog and what the consequences might be.

Remember, dogs do not smile! Showing teeth is generally related to aggressive behaviours.

135

Definitions of animal welfare

Remember that the term **animal welfare** can mean different things to different people.

An RSPCA volunteer

A farmer

We promote animal rights because animals are sentient beings. We need to consider animals because they have feelings just like us, such as pain and pleasure. Good welfare conditions allow animals to experience high levels of pleasure and low levels of suffering.

An animal rights campaigner

Animal welfare means taking positive steps to meet the five welfare needs of animals we are responsible for, and preventing animal suffering.

Animal welfare means providing the right conditions for livestock so that the production rates are good. For example, high egg laying rates are an indication of good welfare.

Dictionary definition
Animal welfare – the proper treatment of animals to ensure a good quality of life.

Links To revise the five animal welfare needs, see page 115.

A scientist

We now know so much more about what animals need. Animal welfare means fulfilling three states of the animal: ensuring that the animal has a good physical and mental state, and is able to fulfil its **natural** needs and desires.

How would you define animal welfare?

An employee from the World Organisation for Animal Health

A veterinary surgeon

Animal welfare means how well an animal is coping with its living conditions. We need to ensure we respect the five animal freedoms, as far as possible, so that we provide the right environment for the animal to cope well.

A welfarist

Welfarism is a utilitarian attitude towards the well-being of non-human animals. We believe that animals can be used by humans, if the cost or suffering to the animal is less than the benefits to humans. The animals we use must be treated humanely.

Animal welfare means ensuring an animal's good health. This involves the prevention and treatment of disease, providing the correct diet and exercise, and ensuring freedom from abuse and exploitation, discomfort and pain.

Links For the Five Freedoms, see page 139.

Links To revise utilitarianism and other ethical theories, see page 132.

Use the definitions above to help you. Think about what each person's priorities are - there will probably be some overlap between the different viewpoints.

Now try this

Explain the conditions needed to ensure good welfare for ducks, from the point of view of:

1 an RSPCA volunteer

2 a farmer

3 a veterinary surgeon.

Animal welfare legislation

You need to be aware of the **legislation** which governs the daily business of organisations and individuals who keep animals in a range of different settings.

Links To revise the five animal welfare needs, see page 115.

Animal welfare legislation in the UK

The main animal welfare laws	The main points of these animal welfare laws
England and Wales: Animal Welfare Act 2006 **Scotland:** Animal Health and Welfare (Scotland) Act 2006 **Northern Ireland:** Welfare of Animals Act (Northern Ireland) 2011	• Cover all **vertebrate animals**. • Apply to **anyone responsible for animals**, including domestic, working and farmed animals. • **Offences** include: causing or allowing an animal to suffer, docking a dog's tail, poisoning an animal, organising animal fights. • Owners/keepers must take **all reasonable steps** to meet the five welfare needs of their animals. • Allow **inspectors** to intervene to improve animal welfare.

Definitions

Terms commonly used in animal welfare legislation include:

– **prevention of harm** – it is a criminal offence to allow an animal protected by the legislation to suffer unnecessarily

– **duty of care** – the legal responsibility of those who own/keep animals to ensure the welfare of the animals they look after

– **an act of his** – something done that has legal significance, usually a criminal act

– **failure of his to act** – a person not doing something that is their legal responsibility.

Legislation revision checklist

There are many other laws governing the welfare of animals. Make sure you learn the key facts about each main piece of legislation:

☑ Which species/situations it applies to.

☑ What responsibilities it requires of someone working with the animals.

☑ Who is involved in enforcing the law, e.g. local authorities, vets, etc.

☑ The year the legislation was passed.

Links Revise the legislation covering animals in different settings on pages 140–150.

Penalties for breaking the law

The penalty for cruelty towards an animal or not providing for an animal's welfare depend on the severity of the offence.

for a period of time or permanently

A ban on owning animals

A fine that can vary between local authorities

Types of penalty

Community service or unpaid work

Imprisonment

Case study: prosecution

Prosecution: brought by the RSPCA in 2015

Defendants: four male students aged 16

Offence: theft and torture of a Chihuahua puppy

Legislation: Animal Welfare Act 2006 section 4

Pleas: guilty

Number of convictions: four

Sentence: All disqualified from keeping animals for five years; one-year youth referral order. Father of one youth issued fine of £5791.50

Remember that the RSPCA brings private prosecutions, but RSPCA inspectors are not themselves law enforcers – they need police assistance to gain entry to a property, for example.

Now try this

Revise the five animal welfare needs on page 115, and apply them to what you know about guinea pigs.

What responsibilities does the owner of a guinea pig have under the Animal Welfare Act 2006?

Role of the government in animal welfare

You need to know about the different bodies that are involved in legislative practice (making and enforcing laws).

Definitions

Government – the team of ministers and peers led by the prime minister – responsible for running the country, developing policies and drafting new laws.

Parliament – all the MPs in the House of Commons and the peers in the House of Lords – responsible for passing new laws and monitoring the work of government.
It considers proposals for new legislation, such as tighter controls on dog racing or greater protection for police dogs and horses when carrying out their work.

Local authority (or local council) – local government – it has many responsibilities, including law enforcement.

Government departments

Government departments are responsible for putting government policy into practice. They are staffed by civil servants and are headed by government ministers. **The Department for Environment, Food and Rural Affairs (DEFRA)** is the main government department responsible for passing and amending animal welfare legislation in the UK.

DEFRA is also responsible for:

- inspections of premises and animals that are registered at farming establishments, for example, cattle and pigs
- tracking and inspecting paperwork and payments for animals registered
- investigating and controlling disease in both plants and animals, for example, foot and mouth disease, and avian flu
- working with the environment, for example, to ensure safety, such as better flood measures.

The role of local authorities

The **local authority** implements some aspects of animal welfare legislation in its area. It can carry out enquiries, visits and inspections to determine whether to issue/renew compulsory licences, for example, to a pet shop, boarding kennels or riding establishment. It might check that:

- the correct species and numbers of animals are being kept
- good levels of welfare are maintained – living conditions, diet, breeding, etc.
- the correct facilities are provided, e.g. exercise area in kennels
- provision has been made for the health and safety of handlers or technicians and the general public.

Links See page 151 for the role of the local authority in specific settings.

The role of the European Union (EU)

While the UK is part of the EU, we are bound by EU legislation.

- EU **directives** are legally binding for all member states, but each state can decide how to incorporate the directives into its national laws.
- EU **regulations** are already law and are binding on all citizens of the EU states.
- EU **decisions** are laws that related to specific cases. They are only binding on the member state(s), organisations or individuals to whom they are directed.
- EU **recommendations** are not legally binding, but allow stakeholders (the people involved) to suggest a course of action.

Now try this

You are considering opening a new pet shop.

1 What is the role of your local authority in your application?
2 What checks might they carry out?

Legislation and codes of practice

You need to be aware of how current animal welfare legislation and codes of practice have developed, and of the differences between the two.

Key events in the development of animal welfare legislation

1965: Brambell Report into welfare of intensively farmed animals. Recommended that animals should have the freedom to stand up, lie down, turn around, groom themselves and stretch their limbs.

▼

'... conditions which appear to us tolerable today may come to be considered intolerable in the future.'

▶

1979: The **Farm Animal Welfare Council** (FAWC) developed the five ideal states of animal welfare, known as the Five Freedoms:

1 Freedom from hunger and thirst

2 Freedom from discomfort

3 Freedom from pain, injury or disease

4 Freedom to express (most) normal behaviour

5 Freedom from fear and distress.

▶

2006: Animal Welfare Act sets out five animal welfare needs:

1 suitable environment

2 suitable diet

3 able to exhibit normal behaviour patterns

4 housed with, or apart from, other animals

5 protected from pain, suffering, injury and disease.

 Links To revise how to meet the five animal welfare needs in practice, see pages 116–130.

Legislation and codes of practice

Legislation sets out the legal requirements for treating animals correctly.

Codes of practice provide guidance on how to comply with legislation. Codes of practice for different types of farmed and domestic species are produced by DEFRA in England, and the governments of Wales, Scotland and Northern Ireland.

Failure to follow the codes of practice is not in itself illegal, but can be used as evidence of how animal welfare legislation may have been broken.

 Links To revise the role of DEFRA, see page 138.

Policies and procedures

Every organisation that works with animals needs to have **policies and procedures** to help employees follow the legislation and codes of practice for the animals in their care. Ethical frameworks can contribute to these. An example is an organisation's policy on when an animal may be euthanised and its procedures for carrying this out.

Links To revise ethical frameworks, see pages 133–134.

Sentience and pain perception

Two factors that have influenced animal welfare legislation are:

1 Scientific evidence that vertebrate animals are **sentient** beings. This means they have feelings and, crucially, they are capable of suffering.

2 Scientific research into **pain perception** has shown that animals feel pain. An animal's physiological and behavioural responses are indicators of pain. Anticipation of pain causes anxiety and fear, and animals seek to avoid pain.

The 3Rs in animal research

The 3Rs are an example of how codes of practice can become law. The Animals (Scientific Procedures) Act 1986 requires researchers to ask these questions before using animals in testing:

☑ **Replacement** – what alternatives are there?

☑ **Reduction** – how can the number of animals involved in testing be reduced?

☑ **Refinement** – how can the correct environment be provided?

Links To revise the legislation governing animals used in research, see page 147.

Now try this

Describe how the Brambell Report influenced the development of animal welfare legislation.

Pet animals

Keeping animals as **pets** is common practice in the UK. Both pet shops and pet owners must comply with the relevant legislation.

Pet Animals Act 1951

This legislation applies to anyone **selling vertebrates** (mammals, fish, birds, reptiles, amphibians) as **pet animals**.

No pet shop can operate without a **licence**. Before it issues a licence, the local authority will carry out an **inspection** or authorise a vet to do this. Inspections can also be carried out at any time after a licence has been granted. The licence must be **renewed** annually.

Penalties for non-compliance

Keeping a pet shop without a licence	Fine and/or prison sentence
Failing to comply with licence conditions	
Obstructing or delaying an inspector	Fine

The pet shop keeper may have their licence cancelled and be disqualified from keeping a pet shop and/or from keeping animals for a specified length of time.

Conditions for obtaining a pet shop licence

Animals must be:

- sold from **permanent premises** – not a market stall or the street
- kept in **suitable accommodation** (adequate size, temperature, lighting, ventilation, cleanliness)
- have suitable **food and drink**
- be **visited** at suitable intervals.

Other conditions:

- Mammals must be **weaned** before being sold (around eight weeks old).
- Precautions for **prevention of disease** must be taken.
- Provision must be made in case of **fire** or **other emergency**.
- Animals must not be sold to children **under 16**.

Pet owners

Pet owners must comply with the Animal Welfare Act 2006. They have a **duty of care** to meet the five welfare needs for their particular animal(s).

Many animals need to have companionship of their own species for their welfare needs to be met.

Ethical considerations for pet shop keepers and pet owners

Which animals is it appropriate to sell and/or keep as pets?

Dogs and cats have been companions of humans for thousands of years. But is it ethical to keep wild animals as pets?

Stakeholders might include the animals, animal owners, business owners and welfare organisations.

Links To revise the Animal Welfare Act 2006, see page 137.

Links See pages 133–134 to revise how to use an ethical framework to help you consider ethical questions like this one.

Links See page 132 to revise ethical theories. The deontological theory might be particularly useful here.

Possible welfare issues for pets

Inappropriate housing/treatment might include:

👎 chaining up animals / leaving animals in hot cars
👎 keeping large dogs that need lots of exercise in a small flat
👎 lack of treatment for illness/injury
👎 overfeeding or feeding the wrong food.

Now try this

List the types of penalties that a court might impose on the owner of a pet shop that does not satisfactorily meet the requirements of the animal welfare inspection and monitoring process.

Dogs

Dogs are a popular pet and are commonly shown at local and national competitions. There are several ways of acquiring a pet dog, including from rescue centres and buying from breeders.

Legislation

Breeding and the sale of dogs are governed by:

1 Breeding of Dogs Act 1973 and 1991

2 Breeding and Sale of Dogs (Welfare) Act 1999.

An establishment must obtain a licence from the local authority in order to breed or sell dogs. It can be inspected at any time. A warrant can be obtained to gain entry to premises where an offence is suspected.

The legislation also applies to people breeding dogs in their own homes if more than five litters are born and sold in a 12-month period.

The legislation covers living conditions and care, and breeding and sale.

Requirements of the legislation

- Suitable accommodation: must have appropriate food, drink, bedding, exercise and regular visits.
- Disease control in place, for example, hygiene and vaccinations.
- Provision for protecting the dogs in case of fire/emergency.
- Bitches under one-year-old must not breed.
- A bitch must not have more than six litters in her lifetime.
- Puppies must not be sold under eight weeks of age (unless to a licensed pet shop).
- Dogs can only be sold at a licensed breeding establishment or licensed pet shop.

Penalties for non-compliance

(X) For no licence or contravening the conditions of licence: a fine and/or imprisonment.

(X) For obstructing or delaying an inspector: a fine.

A breeder can lose their licence and be banned from keeping a dog breeding establishment for a specified time.

The Dangerous Dogs Act 1991

This law **bans owning, breeding or selling dogs bred for fighting**, for example, pit bull terriers and Japanese Tosas. It also makes it an offence for a dog to be seriously out of control. Penalties include **fines** and **prison sentences** from six months to 14 years (depending on whether a dog causes injury or death). The court may order a dog to be **muzzled, neutered or destroyed**.

Ethical considerations

- Should dogs be allowed to breed when there are so many dogs in rescue centres?
- Should anyone be allowed to buy a dog? (Dog breeders do not usually 'vet' new owners, unlike rescue centres.)
- Prevention of inbreeding: The UK Kennel Club's Assured Breeder Scheme promotes good practices to prevent ill health caused by inbreeding.

Links See pages 133–134 to revise using ethical frameworks to address questions like these.

The Microchipping of Dogs (England) Regulations 2015

All dog owners must **microchip puppies by eight weeks old**. The benefits include: reuniting lost dogs with their owners, preventing theft and tracing dangerous dogs.

Now try this

State the requirements under the Breeding of Dogs Act 1973 and 1991 for:

(a) the age at which bitches can be mated

(b) the maximum number of litters that a bitch can have in her lifetime

(c) the minimum age at which a puppy can be sold.

Riding and boarding establishments

Riding establishments keep horses to be let out for riding or to provide riding instruction. **Boarding establishments**, such as kennels and catteries, provide overnight accommodation for animals, particularly for cats and dogs, but other species can also be catered for.

Riding Establishments Act 1964 and 1970

Before granting a licence, the local authority needs to be satisfied that the general welfare requirements for an animal establishment are met (see below).

The council will also need to ensure that the horses:

- ☑ are **suitable** for the purpose they are kept for
- ☑ are **physically fit** and in good health
- ☑ are suitably **groomed**
- ☑ have **well-maintained feet** (trimmed or suitably shod)
- ☑ are adequately **rested**, as well as **exercised**.

The licence holder of the establishment must:

- ☑ keep a **register** of all horses aged three years and under
- ☑ have suitable **insurance**
- ☑ **not** leave someone under the age of 16 in charge
- ☑ **not** hire out heavily pregnant mares or horses under three years old for riding or teaching.

The licence holder or manager must be suitable and qualified to run a riding establishment (having experience with horses or an approved certificate).

🔗 **Links** The penalties for non-compliance with the riding and boarding establishment laws are similar to those for the Pet Animals Act 1951 – see page 140.

Animal Boarding Establishments Act 1963

The licence holder of a boarding establishment must keep a register of all animals received, date of arrival and departure, the name and address of the owner.

They must obtain proof of up-to-date vaccinations before admitting an animal.

Requirements that apply

A boarding or riding establishment must provide for the **general welfare** of the animals, including:

- suitable accommodation for the species, e.g. stables for horses should be at least 3.65 m × 3.65 m
- suitable food, drink, bedding, exercise and visits
- precautions against disease, including isolation facilities
- protection and evacuation of animals in the case of fire.

A licence will not be issued if a person is disqualified from keeping animals or from running an animal establishment.

Ethical considerations

- What are the ethical issues involved in keeping horses for riding and teaching?
- What should boarding establishments do with unclaimed animals or those that become seriously ill during their stay?

🔗 **Links** See pages 133–134 to revise how to use ethical matrices to consider questions like these. See page 132 for ethical theories. Organisations' policies and procedures will also differ. See page 139 to revise this.

Now try this

1. Which piece of legislation is implemented by local authorities for riding establishments?
2. Outline some of the things that a local authority inspector would check when considering whether or not to issue or renew a riding establishment licence.

Farmed animals

There are about 900 million farmed animals raised in the UK each year. The **Animal Welfare Act 2006** applies to all animals, even those that are due to be slaughtered at an early age. There are additional regulations aimed specifically at protecting farm animals.

Legislation

The Welfare of Farmed Animals (England) Regulations 2007 (as amended) specifies the rules for:

- the number of competent staff
- regular staff inspections of the animals
- recording of medicinal treatment and mortalities
- allowing freedom of movement
- appropriate buildings/accommodation
- protection for animals kept outdoors
- equipment properly maintained
- providing wholesome feed and access to water
- correct breeding procedures
- detailed rules for the care of different species.

The Welfare of Animals at Markets Order 1990 covers **correct treatment of animals in markets**: requirements for penning, food and water; care of young animals.

 Revise transporting animals on page 144 and methods of humanely killing animals on page 150.

Enforcing the legislation

- DEFRA produces **Codes of Practice** – guidelines for the welfare of different species – to help farmers comply with the legislation.
- **Government inspectors** can visit farms to check on animal welfare.
- Local authorities may carry out **welfare inspections** following complaints about welfare issues, and are responsible for enforcing The Welfare of Animals at Markets Order 1990.
- Farmers who cause unnecessary suffering to an animal, or fail to kill an animal humanely, can be **fined**, **jailed** or **banned** from owning animals.

Links To revise the role of the treatment of farm animals in the development of animal welfare legislation, see page 139.

Ethical considerations

- **Is it ethical to breed animals for food?** Are animals there for the use of humans as a resource, or should animals have the same rights and level of protection from harm as humans?

Links Revise ethical theories on page 132.

- **What should happen to unwanted animals?** For example, what should be done with male chicks of laying hens and male calves of dairy cows?
- **What types of farming methods are ethical?** For example, intensive farming, free-range and/or organic farming practices.

Links Consider the five needs of animals, listed on page 115.

Animal welfare labelling, such as the 'RSPCA Assured' label, requires high animal welfare standards from farms.

Links Revise ethical frameworks on pages 133–134.

Example: hens

Since 2012, battery laying hens must be housed in 'enriched' or 'colony' cages which allow $600\,cm^2$ per bird. The cages must also provide perching, nesting and scratching opportunities, although these are still very limited compared to those of barn or free-range hens.

Free-range laying hens have opportunities for interaction with many other hens, and plenty of space to search for food and perform other normal behaviours frequently.

Now try this

What is the purpose of The Welfare of Animals at Markets Order 1990?

Consider the potential issues for animals at market.

Transporting animals

Every year, many animals are **transported**, including for economic reasons and for private activities. The transportation of animals for economic reasons (particularly livestock: cattle, sheep, pigs, poultry) is covered by **welfare legislation**.

The Welfare of Animals (Transport) (England) Order 2006

This law (and the laws in Scotland, Wales and Northern Ireland) aims to protect animals during transportation, loading and unloading.

The legislation requires that:

- vehicles must be **clean and suitable** for the species (e.g. size, escape-proof)
- some animals **must not** be transported, e.g. those about to give birth; very young animals; sick or injured animals
- infants must travel **with their mother** if they still need assistance with feeding
- **noise** must be minimised
- there must be **adequate lighting** (e.g. for loading, unloading and inspection)
- correct **paperwork** must be available (e.g. an Animal Transport Certificate).

Personnel must be competent in handling and loading animals, and must work in a way that does not cause distress or suffering.

Ethical considerations

- Is it acceptable to eat some species and not others?
- Does transporting livestock animals cause unnecessary anxiety? (For example, should the animals be humanely slaughtered at their existing location and then transported?)

Reasons for transporting animals

- ✓ Food industry – to market
- ✓ Food industry – to slaughter
- ✓ Pet trade – to a new pet store
- ✓ Pet trade – to a new home
- ✓ To seek veterinary treatment
- ✓ For breeding purposes
- ✓ To add to a new collection (e.g. zoo)
- ✓ Education purposes (e.g. to schools or community settings).

Transporting animals on long journeys

- ✓ The maximum time that animals should be transported for is **eight hours**, unless special authorisation is obtained.
- ✓ Food, water and rest **breaks** should be provided during transportation.
- ✓ There must always be **sufficient bedding**.
- ✓ Moveable panels must be available to create **separate compartments**.
- ✓ There must be provision to maintain/adjust **ventilation** and **temperature**.

Inspections

DEFRA is responsible for granting **transporter authorisation**. Local authorities are responsible for enforcing the rules during transportation.

If an inspector considers that animals are being transported incorrectly, he or she may:

- **serve notice** to rectify the issues
- **stop** the animals being transported
- order the animals to be **unloaded** for food, water and rest before continuing the journey
- in extreme cases, they may order that the animals be **humanely killed**.

The maximum penalty for an offence is an **unlimited fine** and **two years' imprisonment**.

Now try this

Describe what a transporter must do to comply with the legislation for transporting animals.

Wild animals in zoos and private collections

A **zoo** is any establishment where wild animals are kept for exhibition to the public. There are over 500 zoos in the UK, including traditional zoos, safari parks, aquaria and butterfly parks.

Dangerous Wild Animals Act 1976 (DWA)

This legislation is aimed at people who want to own any of the animal species listed in the schedule of this act, which include: red panda, gorilla, giant armadillo, caiman, cobra, wandering spider.

Anyone wishing to keep a dangerous wild animal like a poison dart frog must apply to their local authority for a licence.

The licence holder must:

- **specify** the species and number of animals
- provide **suitable, escape-proof** accommodation
- keep the animal(s) **only** at the premises specified on the licence
- hold **insurance** for any damage or injury caused by the animals.

Local authorities can inspect premises and seize any animals not kept according to the conditions of the licence. Keeping a dangerous wild animal without a licence carries a **fine**.

Zoo Licensing Act 1981

The zoo licensing and inspection system is in place to ensure that zoos provide:

- an **appropriate environment** for each species
- a **high standard** of animal husbandry, veterinary care and nutrition.

Zoos must also:

- **Contribute to conservation**, for example, research or training, breeding programmes, repopulating areas/reintroducing animals to the wild.
- **Contribute to education** about conservation and biodiversity, particularly about the wild animals in the collection.

Local authorities grant **zoo licences** and carry out **inspections**. Government-appointed zoo inspectors assist local authorities by making recommendations, for example, welfare improvements to be made within a specified time scale.

BIAZA (British and Irish Association of Zoos and Aquariums)

BIAZA is a professional body representing the zoo community, as well as a conservation, education and scientific wildlife charity. It follows the European Zoos Directive, which sets out the conservation, education and animal welfare requirements throughout the EU.

BIAZA also works with CITES (Convention on International Trade in Endangered Species of Wild Fauna and Flora) to protect fauna and flora from international trade, to help prevent species becoming endangered or extinct.

Ethical consideration

The stakeholders and some of the ethical considerations here include:

The animals: Can they cope in a captive environment? Can boredom and abnormal behaviours be avoided?

Wildlife organisations, e.g. the Born Free Foundation: Can zoos recreate the complexity of wild environments? How effective is captivity-based conservation?

Zoo owners/employees: zoos provide a livelihood and rewarding work.

Visiting public: enjoyment and education from visiting zoos.

Links See pages 132–134 to revise ethical theories and using ethical frameworks to address issues such as this.

Now try this

Name **four** different species that are listed in the Dangerous Wild Animals Act 1976.

The fur trade

Fur farms have been banned in the UK since 2002, but fur farming still takes place around the world and the trade of animal fur is legal in the UK.

Sources of animal fur

There are two sources:

 fur farms

 animals caught in the wild.

Fur farms are mainly located in Europe, North America and China. There are none in the UK, as fur farms were banned under the Fur Farming (Prohibition) Act 2000 in England and Wales, and similar laws in Scotland and Northern Ireland.

Catching wild animals for the fur trade mainly occurs in North America and Russia.

Fur is either auctioned or privately sold to manufacturers and the finished fur products make their way to retailers internationally.

CITES

CITES stands for the **Convention on International Trade in Endangered Species of Wild Fauna and Flora**. It is an international agreement between governments. Its aim is to ensure that international trade in species does not threaten their survival.

The agreement covers the fur trade. The trade of species threatened with extinction is under strict regulation in order not to endanger the survival of these species further.

Chinchillas

Chinchillas are classed as **endangered** on the IUCN (International Union for Conservation of Nature) Red List of Threatened Species. This means they are at **serious risk of extinction** in the wild. They have been extensively hunted in the past, almost to extinction – one of the reasons for this was the fur trade.

Chinchillas are bred in captivity for fur farms and the pet industry.

Due to their IUCN status, the international trade of chinchillas is closely monitored by CITES and wild-caught trade is prohibited.

Trading in endangered species

CITES monitors the import and export of endangered species by presentation of a permit. These permits are only granted when the Scientific Authority and the Management Authority of the area are happy that the import or export meets the following conditions:

1. The trade will not be detrimental to the **survival** of the species.

2. The animals were **not** obtained in an **illegal** way.

3. The animals will be prepared and shipped in a way that **minimises** the risk of injury, damage to health or cruel treatment.

The fur trade in the UK

Fox and mink are types of fur commonly imported to the UK. There is a ban on imports of furs or skins of endangered species without a permit. There is a total ban on imports of furs from:

- certain species, including seals, cats and dogs
- animals caught in leg-hold traps.

Now try this

On the IUCN Red List of Threatened Species, the status of the American mink is Least Concern.

 You will need to read the whole page for the information you need to answer this question.

According to CITES, what methods can be used to obtain fur from a mink and why?

146

Animals in science and education

Animals are used in **scientific research** and for **educational purposes**. The UK has rigorous welfare regulations for laboratory animals, but this remains a highly controversial area.

Animals (Scientific Procedures) Act 1986 (ASPA)

This law covers vertebrate animals that undergo experiments (called '**regulated procedures**') for scientific or educational purposes which may cause the animals pain, suffering, distress or lasting harm. Animals may only be used in scientific procedures when there is 'no validated alternative and when the potential benefits outweigh the harms'.

 To revise the '3Rs' of use of animals in research, see page 139.

To find out more information about the law on animal research, go to www.gov.uk and search for animal research and testing.

Most animal research is carried out on mice, rats and rodents, with fish and birds making up most of the remainder.

Conditions for using animals in science

- Three licenses must be obtained:

  1 a '**personal licence**' for the researcher

  2 a '**project licence**' – specifies the programme of work

  3 a '**place licence**' – specifies where the procedures will be carried out.

- Qualified personnel must be responsible for the care and welfare of the animals.

- Anaesthetics and pain killers must be used if research procedures cause pain.

- Death as an end-point in procedures must be avoided and replaced with an early and humane end-point, where necessary.

Enforcing ASPA

Government-appointed inspectors carry out checks on research establishments. They may issue notice to improve welfare practices or revoke licences where necessary. Offences carry penalties of fines and/or imprisonment.

Reasons for research on animals

- To learn more about **how the body works**, for example research into pain perception.

   To revise research into pain perception, see page 139.

- To learn **how disease affects the body**, for example, cancer research.

- To develop and test new forms of **treatment** before testing on humans, for example, drug trials.

- To **research genetic manipulation**, for example, to reduce likelihood of developing diseases, or increase desirable traits for food industry.

   For more on genetic manipulation, see pages 33–35.

It is **illegal** in the UK to test cosmetics or household products on animals.

Animals in education

Animals are frequently used as **teaching tools** in schools and colleges, for example, for dissection, behaviour studies, and for animal management and veterinary students. They may be used for education in animal collections such as zoos, for example, in demonstrations. The animals are often used to teach how particular species should be kept and handled.

Possible welfare issues include:

- **overuse** of animals, for example, large classes handling small number of animals

- **stress** of being in a busy environment

- what happens to animals kept in schools or colleges during **holidays**.

Now try this

A college tutor wants to use their own dog to teach a class of 20 students how to take the temperature, pulse and respiration rate of a dog.

What are the ethical considerations in this situation?

You can use an ethical matrix to consider the different stakeholders involved and relevant ethical theories. Refer to pages 132–134 to remind yourself how to do this.

Animals used for entertainment

As animal welfare concerns have developed, legislation has been passed to protect animals used for **entertainment** and certain practices have been banned.

Performing Animals (Regulation) Act 1925

This legislation covers **vertebrate animals used for performance** or in **exhibits** where the public is admitted. It does **not** apply to training animals for the military or police, or for agriculture or sporting purposes.

- **Registration** with the local authority is needed in order to exhibit, use or train animals for public performance.
- The **details** of the animal and the type of performance must be stated.
- Cruelty may result in **loss of registration** or a **ban** from exhibiting or training animals. Exhibiting or training performing animals without registering carries a **fine**.

Performing animals

Circuses: The use of animals in circuses is becoming less frequent because of welfare concerns, for example, over animals being moved around regularly.

Film and TV: Animals tend to be well trained to cope in front of many people and cameras. Species trained to feature in TV and film include:

- domesticated species, for example, dogs, goats and rabbits
- wild animals, for example, tigers, bears and chimpanzees.

Racing animals

Horse racing: The **British Horseracing Authority** produces regulations covering the welfare of racing horses – using a whip during racing falls under these rules.

Dog racing: Welfare issues may include **overbreeding**, and the fate of dogs **not deemed fit** for the industry (for example, lack of drive to chase, not fast enough, or too old to race).

The racing of dogs and horses is often associated with gambling, and the potential to win large sums of money.

Hunting

The **Hunting Act 2004** made it illegal to hunt wild animals with dogs. Before this law was passed, animals such as foxes, deer and hares were commonly hunted in this way. Many people now enjoy **drag hunting** instead, which involves dogs following an artificial trail instead of the scent of hunted animals.

Fighting

Organising **animal fights** such as with dogs and cocks has been banned since 1835, and is an offence under the **Animal Welfare Act 2006**. Illegal fights still do take place, usually for financial gain. These, and other criminal activities involving animals, are monitored and tracked by the RSPCA's special operations unit, who will prosecute offenders.

Ethical considerations

- **Should wild animals be hunted?** Stakeholders might include the public, farmers and those practising hunting, as well as the animals themselves.
- **Should domesticated animals be allowed to hunt?** For example, dogs hunting smaller wild animals.
- **Is it good to see performing animals?** Consider welfare issues of animals no longer required for performances. Compare these with practices in zoos, e.g. conservation of rare species and education.

 Links See pages 132–134 to revise ethical theories and use of ethical frameworks.

Now try this

Explain how you would meet the welfare requirements of a goat being used to perform in TV or a film.

Wildlife and conservation

Wild animals are classified as those that are **not normally domesticated in the UK**. Conservation involves protecting wild animal species and their habitats.

Conservation management

Conservation management programmes aim to maintain **biodiversity** (the variety of animal and plant species within an environment) and ensure they remain in balance. Its primary aim is to **protect the environment**, but also involves **animal welfare issues**. Farmers and game keepers play an important role in environmental conservation. They look after game species, as well as other wildlife and their habitats.

Ethical consideration: Should humans control plant and animal populations, or let nature take its course?

Conservation schemes

These include:

✓ setting aside protected **conservation areas**

✓ **removal** of invasive **non-native** species of plant and/or animal

✓ **reintroduction** of plant and/or animal species to an area

✓ **culling** (pest control).

Links See page 150 to revise reasons for and methods of culling.

Legislation to protect wild animals

These laws have developed for a combination of conservation and animal welfare reasons.

- **Wildlife and Countryside Act 1981:** it is an offence to kill, injure or take any wild bird or to disturb its nest or take its eggs. A wide range of other wild animals (from moths to dolphins) and their shelters are also protected.

- **Wild Mammals (Protection) Act 1996:** prohibits violent acts with intent to cause suffering to wild mammals, but allows humane killing to relieve the suffering of a seriously injured mammal.

- **Countryside and Right of Way Act 2000:** allows public access on foot to open land, while making it an offence to: disturb the nests of specified bird species; disturb specified animals or their shelters; dump non-native species in the wild.

Protection of Badgers Act 1992: it is an offence to capture, injure or kill a badger; to destroy or interfere with an active badger sett; or to allow a dog to enter a badger sett.

Penalties for offences under these laws include a **fine** and/or **prison sentence**.

Veterinary treatment of wild animals

There are ethical considerations regarding treating sick or injured wild animals. Wild animals are not used to being handled by humans and may suffer while in captivity. The **Veterinary Association for Wildlife Management** recommends that the **animal's welfare** should always come first. The animal should be given treatment if it can be released to the wild after recovery. In other cases, **euthanasia** may be more appropriate.

Consider which of the laws outlined above would apply in this case.

Now try this

A fox has been hit by a car and is lying at the side of the road. It has obviously been badly injured, with broken bones and a lot of blood.

Would it be legal for whoever found the fox to kill it?

Killing animals

One of the most difficult ethical issues facing those who work with animals is whether and in what situations it is acceptable to kill animals. You need to know about the different **reasons** why animals may need to be killed and the **legal humane methods** for doing this.

Definitions

☑ **Euthanasia** – killing to relieve an animal from excessive suffering

☑ **Culling** – killing to control population size, prevent spread of disease or to strengthen the breed

☑ **Slaughter** – killing animals for food.

Legislation

The **Welfare of Animals (Slaughter or Killing) Regulations 1995** aim to prevent avoidable suffering during the slaughter process. For example, consideration of the handling techniques used as well as cattle, sheep, goats and pigs must not be able to see other animals being bled. Slaughterhouse inspections are carried out by official vets from the Food Standards Agency to ensure they follow the regulations.

Approved methods of killing animals

With the exception of religious slaughter practices (see below), animals must be **stunned** (rendered unconscious and incapable of feeling pain) before slaughter.

Captive bolt gun and sticking	Electrical	Controlled atmosphere
The **bolt gun** fires a metal bolt into the brain to **stun** the animal. The animal is then very quickly shackled by the hind leg and its throat is cut, severing the major blood vessels and causing rapid death by bleeding out (called **sticking**).	An **electrical current** is passed through the brain to **stun** the animal before sticking or (for poultry) moving on to a mechanical neck cutter. Alternatively, the current may be also passed through the heart to stun and kill the animal simultaneously.	Gases are controlled in a sealed chamber. Animals may be exposed to **anaesthetic** gases first, followed by high concentrations of **carbon dioxide**. The chamber must not be overcrowded and should only house one species.
Species include: cattle, deer, sheep, pigs.	Species include: sheep, poultry, pigs.	Species include: pigs, poultry, wild animals, animals used in research, e.g. rodents.

Euthanasia of suffering pet animals is usually carried out by injecting the animal with an overdose of a sedative drug.

Links For different views on the ethics of killing animals for human use, see ethical theories on page 132.

Culling of animals

Culling is carried out to:

- **control** animal populations (for example, deer culling to avoid overgrazing)

- **prevent the spread of disease** (for example, tuberculosis from badgers to cattle; foot and mouth among farm animals)

- **prevent damage** to **property** or farmland; **prevent damage** to **food** resources (for example, crops or fish supplies).

Religious and cultural attitudes

Some cultures view different species as having higher value than others: some are seen as food resources and others as pets. Cats and dogs are viewed as pets in the UK, but as food sources or vermin in other countries.

Religious slaughtering practices are exempt from pre-stunning the animals. For example, Shechita (Jewish) and Halal (Muslim) slaughter methods involve quickly cutting the jugular vein, carotid artery and windpipe with a sharp knife, without pre-stunning. This method is considered humane and painless by those who practise it.

Links See page 131 for more on religious and cultural views on animals.

Now try this

Explain why animals should be stunned, for example, using a captive bolt gun, before being killed.

Animal welfare inspectors

An **animal welfare inspector** ensures that regulations are being followed, particularly the Animal Welfare Act 2006 and any other regulations that apply to the setting.

The animal welfare inspector

Welfare inspectors are employed to help promote and monitor the welfare of animals.

Responding to welfare issues – complaints made by the public; referrals from the police

Conducting assessments of animals – including follow-up assessments

Rescuing animals in immediate danger – swan tangled in net; cat stuck on a roof

The inspector's role might include

Inspectors may be required to complete animal assessments e.g. checking their health.

Preparing a case file and attending court – if prosecution is necessary

Links To revise the preparation of action plans and case files see page 155.

Creating reports and action plans – to improve the welfare of animals inspected

Educating owners on correct husbandry management – diet, exercise, accommodation, etc

Following legislation – calling police to gain entry to premises

Types of welfare inspector

Inspectors may be employed by the **government**, a **local authority** or **welfare organisation**.

Type of inspector	Examples of settings and types of inspection
RSPCA inspector	Investigates, promotes and monitors welfare, particularly reports of neglect or cruelty; checks on animals that are trapped, beached, etc.
Government inspector	Inspects records (e.g. movement), ID systems (e.g. tagging) and welfare; particularly at establishments where animals are registered, e.g. farms, to ensure all animals in the food industry are controlled.
Qualified vet	May aid a local authority to inspect premises (e.g. pet stores), e.g. to ensure compliance with legislation before issuing/renewing a licence.
Local authority welfare officer	Monitors animal welfare in settings requiring a licence (e.g. boarding establishment or pet shop), to ensure compliance with legislation before issuing or renewing the licence.

Definitions

Objectivity

- Being unbiased or impartial
- Not influenced by other people's opinions
- No personal feelings are involved.

Subjectivity

- Being biased or one-sided
- Other people's opinions can influence decisions
- Personal feelings are involved.

Remaining objective

Avoiding subjectivity can be difficult when witnessing cruelty and neglect. **Compassion** and **empathy** are good traits for inspectors, but they must remain **objective**, as this will achieve better outcomes for an animal's welfare. An inspector who is **unbiased** can remain focused on the **legislative requirements** rather than their own feelings. This will help them to (for example):

- gather evidence for a prosecution in difficult circumstances
- spot where lack of education is likely the reason for poor welfare, rather than wilful neglect.

Now try this

List **five** different duties of an animal welfare inspector.

Be clear and precise and make sure you mention five different aspects of the inspector's role.

Welfare appraisals

Welfare appraisals are **formal** assessments or checks of an animal's health, welfare and condition. The type of appraisal used will depend on the setting and purpose of the appraisal.

Frequency of appraisals

The frequency will vary according to the setting and purpose of the appraisal.

The majority of licence-related appraisals are carried out annually, for example, dangerous wild animal licence, pet shop licence, animal boarding establishment licence or riding establishment licence.

Some appraisals are carried out only when concerns are raised, for example, RSPCA inspections.

> **Links** To revise the welfare legislation that applies in different settings, see pages 140–149.

Formal and informal appraisals

- **Formal appraisals** are normally carried out by a governing body or an animal welfare inspector. They will follow a set procedure and document their findings using appropriate paperwork.

  For example, a farm inspector checking sheep or goats will complete detailed welfare checks of 60 animals, or all the animals if there are fewer than 60.

- **Informal assessments** may take place in some businesses such as pet shops, and would be done by shop staff to check the welfare of their animals.

Carrying out an appraisal

Animal Welfare Appraisal Form

Inspector's name: Date of inspection:

Animal:

Owner's name: Telephone number:

Location:

Nutrition/diet	Yes/no	Owner aware
Is water available?		
Does the animal appear to be well fed?		

Environment

Is the environment appropriate for the animal?

Is the area clean?

Is the area safe?

Does the environment provide appropriate shelter?

Is there a resting area and appropriate bedding?

> This appraisal form is based on checking that the five animal needs are being met.

> **Links** See pages 153–154 to revise different measuring techniques used in welfare appraisals.

Different appraisal systems

Traffic light system	Food certification and assurance schemes	Measures of input and output
Uses traffic light colours to grade any welfare concerns. For example: • red – warning • amber – sufficient • green – no issues	Farmers can volunteer to take part in schemes such as British Lion eggs and Red Tractor, which use animal needs indexes as a quantitative measure of welfare.	Input measures assess the resources put into farming: e.g. feed, housing, substrates, fuel. Output measures assess the animals themselves and their products, e.g. meat, eggs, wool.

> For example, you could assess the animals at your education establishment or at a local zoo.

Now try this

Use the assessment form above (or devise your own) to carry out a welfare appraisal of some animals you have access to.

> **Links** Try using some of the appraisal techniques from pages 153–154.

Measuring behaviour

Behaviour is a good indicator of an animal's welfare. Animal welfare inspectors will **monitor** animal behaviour using techniques such as **ethograms**. These techniques can also be used as part of regular **monitoring** by those who own or care for animals.

Ethograms

1 **Compiling the ethogram:** Observers will **categorise** and **define** the types of behaviour of an animal species observed in particular contexts, for example, location, habitat, weather.

2 **Using the ethogram:** The observer will watch the animal(s) over a length of time (normally several hours or more). They will complete a record sheet to record the number of times a particular behaviour on the ethogram occurs, or the length of time the animal demonstrates the behaviour.

3 **Types of observation:** Either one animal will be observed throughout (a **focal study**) or every animal will be observed at fixed time intervals (a **survey**).

4 **Analsying observations: Activity budgets** can be constructed, which give an overall snapshot of the types of behaviour that were observed during the observation.

Behaviour	Description
Sniffing	Investigating areas with nose to detect scent
Licking	Using tongue to pass over an area
Playing	Using legs and mouth to manipulate objects or interact with others
Grooming	Using mouth area or paws to move fur or clean fur, skin or other areas
Vocalising	Making sounds from the mouth such as barking or whining

Dog behaviours can be grouped and descriptions listed in an ethogram.

Visual observations

An animal welfare appraisal usually includes a **visual observation** to see if the animal(s) are demonstrating any stress behaviours or stereotypes. They may also estimate the body condition score of the animals.

The observer must be familiar with the species to determine whether the animal is behaving normally or abnormally.

Links See page 154 for body condition scoring.

Stress behaviours

Stress behaviours are a useful indicator of animal welfare issues, such as health problems or distress caused by living conditions or ill-treatment.

Stress behaviour	Examples
Vocalisation	Dogs barking excessively or whimpering; cats crying excessively
Bar biting	Pigs, guinea pigs or monkeys biting the bars of their enclosure
Pacing and circling	Lions or wolves walking up and down and wearing a path; bears walking repeatedly round in circles
Crib biting	Horses biting an object such as a stable door and swallowing air
Feather plucking	Parrots pulling out their own feathers
Glass surfing	Reptiles or fish moving back and forth across the glass of the enclosure

Some examples of how animals may exhibit stress.

Links For more on animal welfare appraisal measuring techniques, see page 154.

Now try this

1 Describe **two** behaviours an inspector may look for that indicate poor health or ill-treatment in an animal.
2 Describe what an ethogram is.

Welfare appraisal measuring techniques

You need to know about the **different measuring techniques** that inspectors might use when assessing the health and welfare of animals during an appraisal.

Measuring techniques for assessing animal health and welfare

Score	Condition	Detection of ribs, backbone, and pelvis
1	Emaciated	Obvious
2	Thin	Easily detected with pressure
3	Ideal	Barely felt with firm pressure
4	Fat	None
5	Overly fat	None

Links See page 130 to revise parasites.

Indicators of poor welfare

Signs of **parasites**

Signs of **disease:**
- lethargy
- laboured breathing
- blood/vomit
- discharge e.g. eyes, nose
- body temperature

Poor hygiene:
- excessive urine/faeces.

Signs of **injury:**
- wounds/scarring
- incorrect gait.

Abnormal behaviour

Links To revise behavioural indicators and use of ethograms, see page 153.

Body condition scoring (BCS) – using species-specific charts
- weight too high/low

Genetic selection:
- signs of hereditary health and behaviour traits (e.g. hip dysphasia in a German Shepherd dog)

Fear or stress:
- distress or tense body during handling
- avoidance of humans (proximity or physical contact)
- directed aggression
- cortisol analysis – checking level of cortisol (a stress hormone) in blood or urine

Visual health checks

A **visual inspection** may expose signs of ill health or signs of an animal having been ill-treated, for example, used in illegal fighting or for breeding.

Case study: dogs

Dogs should be monitored in their surroundings to observe how they move and interact with other animals and humans. Observe behaviour as well as gait and posture.

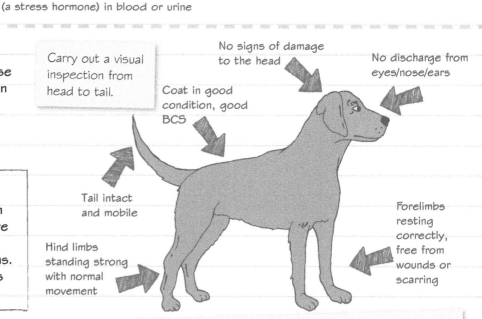

Carry out a visual inspection from head to tail.

No signs of damage to the head

No discharge from eyes/nose/ears

Coat in good condition, good BCS

Tail intact and mobile

Hind limbs standing strong with normal movement

Forelimbs resting correctly, free from wounds or scarring

Body conditioning charts may be used to assess animals raised for the food industry as part of quality assurance schemes.

Now try this

Describe **two** indicators of poor health that an inspector might look for during a welfare appraisal.

Welfare appraisal action plans

Once a welfare appraisal has been carried out, the inspector will need to prepare a report and may need to follow up with an **action plan** or by compiling a **case file**.

Follow-up actions to the welfare appraisal

Welfare appraisal highlights animal welfare issues.

Issues due to lack of education or knowledge about animal care.

Issues due to criminal neglect or cruelty.

Provide expert advice on correct husbandry management for the species.

Produce action plan with prioritised requirements for owners to meet before a certain date.

Decision to prosecute the person or people responsible.

Obtain evidence and create a **case** file to take to court. This will include the findings from visits and appraisals. Evidence may include:
• photographs   • expert advice
• witness statements.
Inspectors have no legal powers to gather evidence themselves, so may require police assistance.

Follow-up visit

Actions met

Actions not met

Depending on the setting, either:
• follow-up inspection
• or no further action
• dependent on organisation.

Negotiate with owners to remove some of the animals (rather than all or none).

Seek police assistance and gain a warrant to enter the premises to remove animals.

Alterations to appraisal structure

There may be changes in the structure of the welfare appraisal, depending on whether it is a **formal** appraisal set by a formal body (e.g. DEFRA) or an **informal** appraisal (e.g the RSPCA).

Reports and action plans

It is important to ensure any appraisal issues are reported appropriately, depending on the situation and type of appraisal being carried out.

If actions are required, they need to be stated clearly and a time scale set for when the actions need to be carried out. The actions may be prioritised according to which need completing most urgently.

Welfare appraisal scenario

An inspector follows up on a complaint from a member of the public that her neighbour's home is overrun with dogs that the neighbour cannot take care of properly. The inspector finds over 20 dogs in poor living conditions, but in reasonable health. The inspector educates the owner on the required conditions and will review with a follow-up visit over the next few days.

Animal hoarding or overcrowding is one of the welfare issues that inspectors might come across.

Now try this

Read the **welfare appraisal scenario** above. On the next visit, the inspector finds that the living conditions of the animals have not improved.

What would the next step in the welfare appraisal process be?

Your Unit 3 set task

Unit 3 will be assessed through a task, which will be set by Pearson. In this assessed task, you will need to carry out research and answer questions based on a set task brief to show your understanding of **animal welfare and ethics**.

Set task skills

Your assessed task could cover any of the essential content in the unit. You can revise the unit content in this Revision Guide. This skills section is designed to **revise skills** that might be needed in your assessed task. The section uses selected content and outcomes to provide an example of ways of applying your skills.

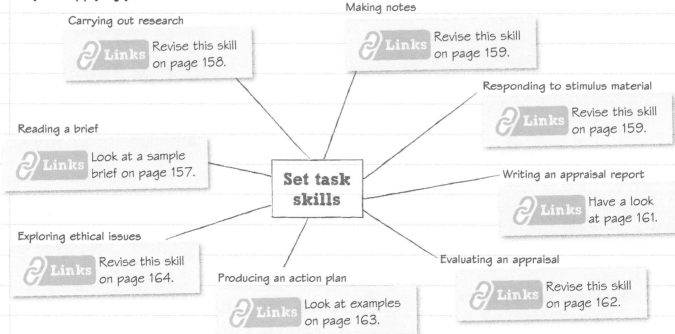

Carrying out research

Links Revise this skill on page 158.

Making notes

Links Revise this skill on page 159.

Reading a brief

Links Look at a sample brief on page 157.

Responding to stimulus material

Links Revise this skill on page 159.

Set task skills

Writing an appraisal report

Links Have a look at page 161.

Exploring ethical issues

Links Revise this skill on page 164.

Producing an action plan

Links Look at examples on page 163.

Evaluating an appraisal

Links Revise this skill on page 162.

Check the Pearson website

The activities and sample response extracts in this section are provided to help you to revise content and skills. Ask your tutor or check the Pearson website for the most up-to-date **Sample Assessment Material** and **Mark Scheme** to get an indication of the structure of your actual assessed task and what this requires of you. The details of the actual assessed task may change so always make sure you are up to date.

You should make sure you take everything you need into your assessment, including a black pen you find easy to use and at least one spare pen.

Now try this

Visit the Pearson website and find the page containing the course materials for BTEC National Animal Management. Look at the latest Unit 3 Sample Assessment Material for an indication of:

- the structure of your set task, and whether it is divided into parts
- how much time you are allowed for the task, or different parts of the task
- what briefing or stimulus material might be provided to you
- any notes you might have to make and whether you are allowed to take selected notes into your supervised assessment
- the questions you are required to answer and how to format your responses.

Researching a task brief

On this page you are given an example task brief and some example task information to help you practise the skills involved in reading and starting to plan your research in preparation for your actual set task.

Revision task brief

You are required to carry out research into the scenario provided in the task information below. You should consider the following areas in relation to the scenario:

- legislation and regulations relating to the animal species

- policies and practices relating to the setting and linkages to the welfare requirements of that species.

Revision task information

Ridgemoor Boarding Kennels in Westshire has recently been inspected by the local authority and found to be unsuitable. Ridgemoor Boarding Kennels has appointed a new manager to review and improve the situation.

You have been asked by the new manager to assist in the appraisal of Ridgemoor Boarding Kennels.

Your notes should be structured appropriately for the brief your are given. You should think carefully about which areas to research based on the brief you are given.

When you get your task brief and information, read through and think about the types of information you need to know for each point.

For legislation, ask questions such as: Which legislation applies to the setting/species in the scenario?

Are licences required?

Are suitable experience or qualifications required?

What are the welfare requirements set out in the legislation?

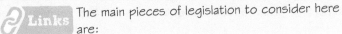 **Links** The main pieces of legislation to consider here are:
- Animal Welfare Act 2006 – see page 137.
- Animal Boarding Establishments Act 1963 – see page 142.

Using good sources

Make sure you use reliable sources when doing your research. Keep a record of useful sources so you can return to them later if you need to. You could use:

☑ books: remember to use the most up-to-date editions

☑ journals: again, check the dates

☑ your course notes and handouts

☑ websites: ensure they are legitimate. In this case, it is advisable to use the actual government legislation webpages.

☑ reports: especially those published by DEFRA or animal organisations (e.g. RSPCA)

☑ online libraries: your centre may have a subscription to an online library

☑ Google Scholar: allows you to search online academic articles.

Show your skills

This task brief is used as an example to show the skills you need. The content of a task will be different each year and the format may be different. Check with your tutor or the up-to-date Sample Assessment Material on the Pearson website for details.

Now try this

Read the revision task brief and task information given above.

1 Highlight the key parts which you need to research.

2 Create a mind map or bullet points for what you would need to research.

Links See pages 158–159 for an example of how you could do this.

Carrying out research

Here are some examples of skills involved if carrying out research.

Good research techniques

☑ Plan what you are researching.

☑ Write a list of the sources (websites, books and other resources) you use.

☑ Allow enough time to create a plan and to write up your notes.

☑ Make sure your sources are:

- **relevant** – check that the source gives you the information you need. Don't include information just because it looks detailed or that you know by heart already.
- **current** – legislation changes and is updated. Make sure you use the most current one.
- **reliable** – consider whether each website you look at contains accurate information.

☑ If you get stuck, go back to your research plan to check what you have done so far and what you still need to do.

☑ It's easy to go off topic, so always refer back to the task brief to prevent this.

Writing research notes

Initially when you are carrying out your research, you can make extensive and detailed notes if you find this helpful, but bear in mind that you may need to then condense these down if you are allowed to take a defined number of pages of notes into your set task assessment. Below you will find an example extract of some initial student notes for the revision task brief given on page 157.

Sample notes extract

Section: Dog welfare in kennels

Many dogs find kennel life challenging – this can influence their welfare. An inadequate kennel environment can make it difficult for a dog to cope.

Examples: too small kennels and restricted exercise can affect a dog's behaviour patterns and compromise welfare.

Animal Boarding Establishments Act 1963 – focuses on physical environment. However, studies show that husbandry regimes and human contact are just as important for dogs' welfare in kennels.

Exercise also important: for physical fitness and to prevent abnormal behaviours. Recent studies on kennelled working dogs showed that dogs which are exercised for less than 30 minutes per day exhibit more repetitive and stereotypic behaviours, bark more and visit vet more frequently.

Dogs must be exercised away from kennel at least once a day, for at least 30 minutes, unless vet advice suggests otherwise.

Safe and suitable toys, bones or other items should be provided to allow the dogs to perform natural chewing behaviour. Check daily and replace if unsafe.

Ensuring good welfare in kennels benefits dogs and kennel staff, and makes business sense. If the dogs have good welfare and don't become stressed, owners are more likely to use the services again.

> Naming specific legislation is helpful – however, you should make sure that you either remember what this covers or expand the notes to include it. This student has referred to 'studies' but not given details of what these are. Good notes would include references to sources.

> Use evidence from journals and studies to back up your points. In your assessment, you must make sure you write in your own words to avoid plagiarism. Only copy text if you wish to quote from a study or journal. Put the quotation in speech marks and include the reference to the sources in brackets after it.

> You need to be unbiased throughout your research and consider the situation from all points of view.

Now try this

Remember to link it to the task brief given on page 157.

🔗 **Links** Meeting the five welfare needs of animals is covered on pages 116–130.

Continue these notes by researching the Animal Boarding Establishments Act 1963.

🔗 **Links** To revise the Animal Boarding Establishments Act 1963, see page 142

Making notes

If you are able to take notes into your supervised assessment, you may be limited in how many pages you are allowed. You should therefore ensure that your notes are appropriate and of a suitable length.

Good notes

☑ Practise different methods of note taking – bullet lists, spider diagrams, flow charts, tables, labelled diagrams. Find out which methods you prefer for different types of information.

☑ Use headings – so you can quickly find the section you need.

☑ Check your notes – are they legible? Do they make sense? Do you understand any acronyms or specialist terms you have used?

☑ Get straight to the point – only include the key points; you don't need full sentences.

☑ Refer back to the brief – is all the information in your notes relevant? Have you included everything you need to?

☑ Do not include information you already know – concentrate on areas you are not so sure of.

Sample notes extract

Inadequate environment (e.g. too small)   Not exercised enough

Dog welfare in kennels

Not given enough or any enrichment (e.g. chew toys)   Not enough human interaction

This is an extract from a student's final notes, based on the task brief from page 157 and the research from page 158.

Use methods such as mind maps, bullet points and short sentences to summarise your notes. These methods will take up less space and be easier to refer to.

Exercise requirements in kennels

Dogs need:

- exercise away from the kennel at least once a day
- ideally exercise for at least 30 minutes
- safe and suitable toys and bones to allow them to perform natural behaviour of chewing
- interaction with humans, other than when carrying out husbandry activities.

Good welfare → happy dog owners, kennel staff, kennel owner and dog is not stressed.

Preparatory notes

You may be allowed to take some of your preparatory notes into your supervised assessment time. If so, there may be restrictions on the length and type of notes that are allowed. Check with your tutor or look at the most up-to-date Sample Assessment Material on the Pearson website for information.

 Links To revise welfare requirements within the animal industry, see pages 116–130.

Now try this

Look at the rough notes you made about the Animal Boarding Establishments Act 1963 for the 'Now try this activity' on page 157. Write up a final version of these notes.

This is only one of the sections which you would cover in your final notes, so it only needs to be a couple of paragraphs.

 Links To revise the Animal Boarding Establishments Act 1963, see page 142.

Responding to stimulus material

As part of your set task assessment you may be given some additional stimulus material. You need to make sure you use this within your responses.

Questions to ask

If provided with stimulus material, you need to read it carefully and assess what is and isn't provided. The example below gives some ideas of the types of question you could ask yourself about the documents provided.

> Remember that the particular documents you could be given will vary.

Example documents in the stimulus material	Example questions you can ask about these documents to help with your appraisal
A floor plan of the establishment	• Dimensions – enclosures big enough? Space for animals to exercise? • Layout – locations of rooms/enclosures appropriate? Social/solitary needs of animals catered for? Provision for emergency evacuation? Feed stored correctly?
A list of animals (e.g. species, numbers, notes on particular animals)	• Are animals being housed with/near compatible species? • Do these animals have any special requirements? • Is a licence needed to keep, sell or breed any of the animals?
Examples of paperwork/ forms, e.g. when selling or admitting animals	• Does the form allow for all the necessary information to be gathered and recorded? • If the form has been filled in, has this been done correctly?
Information on the species/breeds kept for use by staff or customers	• Is the form designed so that it covered all information required? • Have all five animal needs been covered? Are there any diagrams? If so, are they correct?
Notes written by staff	Do these show examples of good practice? If not, how should things be done correctly? (For example, maintenance issues recorded in a book, rather than on sticky notes.)
Staff procedures document	• Are these the correct procedures to comply with legislation? • Are they adequate to ensure the welfare of the species kept? • Is anything important missing?
Guidance from the local council	Read this carefully. It contains information on how to comply with the legislation covering this type of establishment. Use it to help you check whether the establishment meets its legal obligations.

Now try this

Choose **one** of the items of stimulus material provided in the Sample Assessment Material provided by Pearson. Write a minimum of **two** questions you can consider when responding to the material.

Writing an appraisal report

You may be asked to write an appraisal report on the setting in the scenario. Base your appraisal on the information given in the scenario and on any stimulus material provided. Below are some things to remember when writing appraisal reports, along with a sample response from one student writing an appraisal report based on the example task brief given on page 157.

Measuring techniques used to assess welfare

These include:

- ✓ animal needs (measure of the five animal welfare needs)
- ✓ traffic lights system
- ✓ the difference between input (resource) and output (animal) measures
- ✓ assessment by health and physiology
- ✓ assessment by behaviour
- ✓ measuring pain and stress in animals
- ✓ ethograms to create an activity budget.

🔗 **Links** To revise welfare appraisals, see pages 152–155.

You need to:

- make sure your appraisal is to the point, unbiased, clear and covers all areas required
- refer back to the task brief and information – make sure you know who the appraisal is for, in this case, the kennel manager
- consider the quality of care provided at the kennels in relation to the five welfare needs
- consider the legislative and welfare issues at the kennels.

Sample response extract

Welfare appraisal report: Ridgemoor Boarding Kennels

1 Need for a suitable diet: all dogs at the kennels at the time of the visit looked to be getting a nutritionally balanced diet. This was evidenced by carrying out a thorough health check on 5 of the 20 dogs present and visually checking the remainder. Evidence from the stimulus material showed that staff use a whiteboard to monitor and record the feeding of all dogs. Dogs are fed food provided either by the owner or by the kennels, according to the owner's wishes. The food prep. area was clean and well organised.

These strengths demonstrate good practice in ensuring the welfare of the dogs according to the Animal Welfare Act 2006.

However, a few areas need to be improved. Some dogs were fed and then exercised straight after finishing their meal, which is not good for the digestive system and can cause health issues. It is recommended to allow at least 30 minutes after feeding before exercising, so the kennels needs to adapt their daily plan to allow for this ...

This is a good response because the student has used relevant examples drawn from the task information to discuss welfare needs. Your examples should be drawn from or be relevant to the stimulus material provided.

This is just part of the appraisal, focusing on the diet of the dogs.

This is a good link to the relevant welfare legislation and policies.

The student has started to identify the strengths and weaknesses in a balanced way, and rank them in terms of their impact.

However, you are not asked to create an action plan in this activity. This student needs to be careful to focus on the appraisal, not recommendations for change.

This is only part of the appraisal. A full answer would need to go on to cover other legislative and welfare issues at the kennels.

Now try this

Continue with the appraisal above using the information you gathered on the Animal Boarding Establishments Act 1963 in the 'Now try this activity' on page 157. Write a sentence on how this law links to the need for a suitable diet.

🔗 **Links** To revise the Animal Boarding Establishments Act 1963, see page 142.

Evaluating information supplied

You need to be able to evaluate any information supplied to you for the appraisal. To do well in this type of question, your answer should show a comprehensive evaluation of **how suitable** the information supplied is, it should **identify what is missing** and should give **clear reasons** for what you say.

Planning your evaluation

- Allocate a sensible amount of time for the activity based on the marks available.
- Start by looking at any stimulus material provided and listing what else you might need.
- Make sure you answer the question according to any instructions given. There may be more space than you need.

Read the extract from a sample student response below to understand how you might start evaluating the information supplied for the appraisal.

Evaluating information provided

Remember that even if you have considered a variety of information supplied, a true appraisal cannot be carried out without inspecting the premises and the animals.

To evaluate an appraisal, you should:

 Evaluate the information provided.

 Identify relevant additional information that would considerably improve the appraisal.

Sample response extract

A full appraisal cannot be carried out without seeing the premises and the animals that are present there. This is because you won't be able to conduct a health check or see the animal's behaviours. A true evaluation of the animals' health and welfare cannot be determined from the information provided. Video evidence could be used, but this will not demonstrate accurate happenings and welfare standards at the kennels and a false appraisal conclusion could be drawn.

It is important to know and see certain information, such as the staff daily procedures, as you are unlikely to see a full working day while conducting an appraisal inspection. The staff daily procedures will show you how the kennels are run and while you are there you can see part of the daily procedure in action, which will back up the evidence provided and give a true indication of the dogs' welfare for the appraisal ...

The student has started by highlighting generally the **suitability** of the information provided in the stimulus materials. They should then move on to assess the specific information provided.

Explain how the additional information provided would improve the appraisal. A comprehensive answer that explains why each piece of information is important will improve the quality of the response.

 To revise welfare appraisals, see pages 152–155.

Now try this

According to the Animal Boarding Establishments Act 1963, what do you need to see (in person or on paper) to determine whether a boarding kennel is obeying the legislation?

Use the notes you created in the 'Now try this' activities on pages 158 and 159. Think about what criteria the kennels must meet in order to comply with the legislation.

 To revise the Animal Boarding Establishments Act 1963, see page 142.

Producing an action plan

Here are some examples of skills involved if producing an action plan. Read the extract from the sample student response below to see how you could start to structure an action plan for the example task brief given on page 157.

Producing an action plan

An action plan is an important part of the appraisal process, as it clearly states which areas need to be improved and how this can be done. You can present your action plan using a table or under subheadings.

SMART targets are a great way to present actions. See the second sample response below.

Your action plan must include:

- ✓ actions required
- ✓ reasons for the actions
- ✓ timescales for the actions to be completed.

> **Links** To revise welfare appraisal action plans, see page 155.

> This extract covers just one action in the action plan. Make sure your action plan addresses all the issues and factors you identified in your appraisal.

Sample response extract

Action plan for Ridgemoor Boarding Kennels

The areas identified in the appraisal that need addressing are:

1. Amendment of Daily Plan

Certain dogs are being walked shortly after being fed, which can cause issues to their digestive system and welfare.

It is recommended to amend the daily plan, to allow at least 30 minutes after feeding before exercising. This could be done by moving certain cleaning jobs to straight after feeding rather than after walking. This will allow the dogs to eat and start the digestive process off correctly. This will not interrupt the workflow of the kennels, and will allow for all jobs to be completed in a timely manner. This action is to be implemented immediately, together with training of staff in new plan and reason for change ...

> The response would be better if the actions were clearly prioritised and had a timescale for action to show what is most urgent and needs to be done first.

> Make sure you set a realistic timescale for each action. In this case, it is realistic for the kennels to reorder activities straight away. Other action points might need a little more time to implement.

Improved response extract

> Here is the same action but presented using SMART objectives.

Issue to address	Daily Plan does not allow enough time for dogs to digest meals.
Action	Allow dogs at least 30 minutes after feeding before exercising.
How	Move certain cleaning jobs to straight after feeding, rather than after walking the dogs.
Why	If dogs are exercised shortly after eating, this can affect digestion of the food and cause health and welfare issues.
Timescale/ urgency	To be implemented immediately and training put in place for staff, so they are aware of reasons behind change.

Specific

Measurable

Achievable   **R**ealistic

> The student has made sure they have given a reason for the action.

Time-based

Now try this

It was identified in the appraisal that Ridgemoor Boarding Kennels requires maintenance work in kennels number 3 and 14 as the drain covers have broken.

Create an action plan for this issue.

> **Links** To revise welfare appraisal action plans, see page 155.

Exploring ethical issues

Here are some examples of skills involved if asked to explore the ethical issues involved in a given scenario. Read the extract from the sample student response below to see how you could discuss ethical issues.

Exploring ethical issues

Ethics cover a range of situations and human behaviours and can be influenced by factors such as:

- ✓ accepted social practices
- ✓ personal feelings
- ✓ religious beliefs
- ✓ legislation
- ✓ science.

Ensure you are unbiased and cover each issue from different ethical viewpoints.

This student is discussing the ethical issues involved in the owners of three German shepherd dogs choosing to attend Crufts to show their dogs, hoping to achieve best in breed. They have started well by making clear which aspect they are focussing on and then structuring different points paragraph by paragraph.

Sample response extract

There are many ethical issues associated with showing animals in competitions. For the purposes of this activity, I am solely going to concentrate on showing dogs, although many of the points made will also apply to different species that are shown.

The first point to highlight is the welfare of the dogs during the show. Shows are generally day-long events and can mean dogs being kept in a crate, or kept on a lead for long periods of time. Normally they are given opportunities to urinate and defecate. Behaviourists would say that not being allowed to walk freely for long periods is not a natural behaviour for a dog, and could result in stereotypic behaviours being shown.

This practice also contravenes legislation, as one of the five animal needs stated in the Animal Welfare Act 2006, is the 'Need to exhibit natural behaviours'. Animal rights activists might argue that dog showing is solely for the purpose and enjoyment of the owner/breeder, that the dog does not care if they meet breed standards and they do not enjoy being restricted for long periods ...

Ensure you discuss the ethical issues from several different points of view, and, where possible, explain how these viewpoints are interrelated.

Give reasons to support your points and link to relevant evidence.

Use appropriate terminology where possible.

Make sure your answer is relevant to the scenario and support this by giving examples.

Remember that you could use an ethical matrix to help you explore different ethical questions from the viewpoint of different ethical theories and the different stakeholders. See pages 133–134 to remind yourself how to do this.

This is just part of the response. You would be expected to discuss several issues and finish with a judgement and your reasons for this.

Now try this

The sample answer above is just part of a complete answer.

There are many ethical issues that could be discussed regarding showing dogs in competitions. List the other issues that you could include in your answer.

You are likely to come up with lots of different issues. It would be better to discuss three or four different issues in depth and from different points of view, rather than lots of issues only briefly.

Remember that you are being assessed on your knowledge **and** your understanding.

Answers

Unit 1 Animal Breeding and Genetics

1 Reasons for breeding

The majority of animals are bred for commercial purposes in the food industry for human and animal consumption. In addition, many animals are bred as pets, for sport and for conservation.

2 Genetic terminology

Animals have two copies of each gene. For a recessive phenotype to be shown, both genes need to have the recessive allele. If even one dominant allele is present, the animal will have the dominant phenotype.

3 Mendelian genetics and monohybrid crosses

Bb and bb.
Amelanistic is a recessive trait, so offspring need to be homozygous recessive to show this phenotype.

	B	b
B	BB	Bb
B	BB	Bb

	B	b
B	BB	Bb
b	Bb	**bb**

	B	b
b	Bb	**bb**
b	Bb	**bb**

Of the possible combinations, only Bb and bb mates have the potential to produce homozygous recessive offspring.

4 Dihybrid crosses

1 Ewe: Wp, Wp, Wp, Wp
   Tup: wP, wp, wP, wp

		Tup			
		wP	wp	wP	wp
Ewe	Wp	WwPp	Wwpp	WwPp	Wwpp
	Wp	WwPp	Wwpp	WwPp	Wwpp
	Wp	WwPp	Wwpp	WwPp	Wwpp
	Wp	WwPp	Wwpp	WwPp	Wwpp

2 Phenotypic ratio:
white wool, horned = 8
white wool, polled = 8
coloured wool, horned = 0
coloured wool, polled = 0

5 Gene interactions

Incomplete dominance is where neither allele is completely dominant over the other and an intermediate phenotype results.

6 Sources of variation

Crossing over occurs between homologous chromosomes, leading to the independent assortment of genes. This makes it unlikely that two gametes will have the same DNA.

7 Selective breeding

Negative assortive mating, it may increase the variation between offspring in future generations.

8 Desirable characteristics

1 Animals that are well socialised and are more docile make good breeders as they show less aggressive behaviours to other animals and people too. This makes mixing groups together easier and safer.
2 Animals that have unusual markings or colouration are often more desirable to buyers, and can sell at higher prices.

9 Construction of breeding programmes

Animal selection can improve animal welfare by reducing the potential for future generations to inherit genetic disorders or health issues from their parents. Benefits to humans can include higher productivity and disease resistance, which will lead to commercial gain.

10 Factors affecting selection for breeding

A farmer would look at data with regards to milk production and previous breeding. A farmer would want to be aware of whether the heifer has successfully bred in the past. The heifer would be easy to handle and of good temperament. They would have udders and teats of good size and condition.

11 Use of information in animal evaluation

1 Breed profiles are useful as they can provide an overview of characteristics that can indicate whether animals would be suitable or not, for example, as a pet.
2 Information that should be included would be life span, size and temperament, for example. This could indicate whether animals could live together in harmony, allow correct enclosure/accommodation sizes to be acquired, and also consider owner lifestyles.

12 Animal condition

The diet can impact on teeth condition as softer foods such as pouched wet meats can often make teeth decay quicker compared to dried complete mixes. An animal's teeth will also change over time, and older animals will see more 'wear and tear' than juvenile or adult animals. This, along with illnesses such as dental diseases, may cause tooth loss (or further decay) if they are not maintained, or provided with additional care from owners.

13 Behaviour assessment

1 Behaviour assessments establish whether animals can be suitably mixed with other animals as well as work well with handlers. This will ensure the welfare, health and safety of all involved in the breeding process.
2 Undesirable behaviours that may be exhibited from potential candidates could include:
   • aggressive behaviour towards other animals or humans
   • lack of interest in being social with the same species
   • previous breeding demonstrating poor parental care.

14 Legislative requirements for breeding animals

Suitable environment – correct to the purpose and welfare of the animals in a breeding programme (e.g. whelping box of sufficient size for mother and offspring).

Suitable diet – correct to the requirements of the animals in a breeding programme (e.g. age or status – lactating).

Exhibit normal behaviour patterns – animals to be provided with enrichment (e.g. mice given bedding material to nest with, tunnels to hide in).

Housed with, or apart from, other animals – social animals to be kept in groups; solitary animals only to be brought together for mating purposes.

Protected from pain, suffering, injury and disease – all enclosures including areas used for mating to be safe. Animals not be put in stressful situations, which may impact on the animals selected for a breeding programme.

15 Legislative requirements for breeding dogs

Bitches used to breed must be of the correct age (over one year) and only allowed to breed a specific number of times. The bitch can breed once a year (have one litter per year, but a maximum of six in her lifetime).

16 Handling techniques and strategies

Advantages: 'scooping' animals can be very comfortable for the animal as their body weight is supported; it is also an easy technique to implement, needing no equipment or preparation.

Disadvantages: 'scooping' animals offers no security, and escapes are still possible (e.g. mice jumping out of hands); it may also not be ideal for smaller species, e.g. small amphibians, as you cannot see the animal unless the hand is opened up widely (which would allow for escaping).

17 Preparation for breeding

Humboldt penguins are assessed on their ability to perform auditory displays. The male will dig a nest then sit on it and bray; this is called an 'ecstatic' display. Once a mate has been identified, the braying continues with the female joining in, a 'mutual' display. Penguins also use visual cues to identify their mate once the bond has begun to form. Potential mates should be housed where they can hear each other's vocalisations and see each other to allow the mate recognition strategies to be utilised.

18 Reproductive technologies

sperm sexing

19 Contraception

Animals are conscious throughout the entire procedure; the band is applied and there will be significant pressure. The band must be applied for a specific amount of time, and could vary between animals. Animals will need to be restrained too, perhaps in a crush, and therefore experience further anxiety.

Main considerations – stress and pain, however, there is no invasive surgery required, or the use of general anaesthetic.

20 Pregnancy diagnosis in mammals

Any three from:
- increase in weight
- inactivity or restlessness
- increase in teat size
- change in appetite
- dog does not return to oestrous
- vaginal discharge
- distended abdomen.

21 Preparing for parturition

Preparations include:
- preparation of location and accommodation
- gathering of equipment
- signs of parturition
- signs of problems
- preparation and use of PPE
- risk assessments in place.

22 Parturition

It is important to have emergency telephone numbers to hand if there are complications. Additional heat sources may be required, such as lamps, towels. PPE needs to be available if intervention is required. Disposable PPE such as gloves and aprons are often useful.

23 Housing of breeding animals

It can prevent injuries to the handlers by knowing whether to enter the enclosure, or handle the offspring. It can also help assess enrichment types and housing design.

24 Care plans

This takes into consideration age and life stage, whereby their diet changes, it ensures that the animal is monitored, and the correct diet is provided to ensure good health.

25 Care plans: dogs – parturition and neonatal care

Pre-farrowing: prepare accommodation for parturition (farrowing) by disinfecting and drying areas, 'straw up' and provide sufficient bedding. Check heaters and safety of area, as well as drinkers working, clean troughs, etc. Handle sows gently and put in stalls. **Farrowing:** monitor birth process, usually lasts around 2–4 hours, with 10–45 minutes between piglets. Observe for birthing problems and liaise with vet if problems occur. **Neonatal care:** ensure piglets are suckling, fostering may be required. Observe piglets regularly to ensure they are accessing sufficient milk. Provide a clean and dry area, as well as a heated area.

26 Care plans: dogs – early life to weaning

0–3 weeks: Provide milk formula (specially produced for puppies) every 2–3 hours.

3–5 weeks: Introduce a small amount of moistened softened food to accompany milk formula, reducing the milk to 2–4 times a day

5–8 weeks: Increase moistened soft feed, as well as introducing 'harder' feed. Milk should be decreasing throughout the weeks (if still necessary).

8 weeks+: Puppy should be eating harder feed, decrease moistened feeds, as well as cease providing milk. Puppy should be fully weaned.

27 Care plans: poultry – egg incubation

Do not attempt to hatch heavily soiled eggs, disinfect and dry incubator between batches ensuring all egg debris and feathers are removed, and ensure a good supply of fresh air throughout incubation.

28 Care plans: poultry – hatching to fully grown

Provide a variety of environmental enrichment, increase the amount of space per animal, avoid overcrowding, improve husbandry and increase food availability.

29 Care plans: reptiles

Attempting to separate the eggs can damage them, leading to death of the embryos.

30 Congenital and hereditary conditions

Polydactyly is a condition which leads to additional digits in one or more limb. In dogs, this can increase their grip on surfaces, improving their ability to climb, scramble and grip on ice. This would be of benefit in working dogs who are required to perform any of these tasks.

31 Analysing DNA

A sample of cells will be taken from each puppy and the DNA extracted. The sample will be amplified using polymerase chain reaction and restriction enzymes will be used to cut the DNA at specific sequences. These fragments will then undergo gel electrophoresis and analysed to identify the presence of the gene causing ciliary dyskinesia.

32 Genetic screening and disorders

Von Willebrand disease is a genetically inherited condition. Gene testing can be used to determine if a dog has the gene for this disease (is a carrier). This will allow the breeder to make an informed choice regarding if that animal should be used to breed, and, if so, with which animals, to reduce the chance of puppies having the disease.

33 Gene modification

Pharming – using animal products to produce a substance of pharmaceutical value.

34 Genetic modification process

Example answer:
Advantages of using insulin from microbes include the absence of animal welfare issues by using a non-animal source, reduced chance of a bad reaction to the insulin if it was from a different animal species and the potential to increase insulin production easily. Disadvantages include the potential risks to health and the environment of using genetically modified materials.

35 Creating a transgenic animal

Only the cells the DNA is inserted into will be able to produce more cells containing the modified DNA. If the DNA was inserted into an adult ewe, the DNA would need inserting into millions of cells. By inserting it into a fertilized egg, only one cell needs modifying for the change to be incorporated in the organism.

36 Implications of genetic manipulation

An advantage of Enviropigs would be the reduction in pollution caused by the reduced phosphorus content of their faeces. A disadvantage would be the welfare implications of causing the major metabolic changes which would be needed to increase the phosphorus metabolism.

37 Regulation and ethics of genetic manipulation

1 An advantage of this approach would be the increased commercial gain, while a disadvantage would be the negative effect the rapid growth could have on health.
2 Three ethical implications include: the impact the modified salmon could have on the environment, the potential for the modified genes to become part of the wild salmon population, and the impact of the gene modification on the welfare of the salmon.

38 Your Unit 1 set task

Please use the latest information from the Pearson website to answer this question.

39 Reading the brief

Your answer should include all the sections listed in the sample responses given on page 37.

40 Carrying out research

Your notes might include the following:
- **Females** – increase in appetite (to support egg development).
- **Courtship behaviour** – the males' throats (beards) darken to almost black. Males exhibit head bobbing and show more interest in the females. Females respond to head bobbing with similar motions, along with waving-type behaviour – this is a signal to the males that female is ready to mate.

- **Copulation** – male chases female, climbs on her back and bites to hold in place. This happens several times over a period of a month.
- **Egg laying** – 4–6 weeks after copulation and gravid (carrying eggs), females become plump before laying. Female feeding habits slow down and she may stop eating for a few days prior to laying. Female searches for a safe place to lay her eggs, starts digging in a substrate and buries them.

41 Making notes

Your notes should cover several different points of view, including personal feeling, legislation, health concerns, animal rights and welfare, etc. They might include the following:
- Should we keep reptiles or any other animal as a pet?
- Personal gain is for us only.
- Animals are stressed due to living in captivity.
- A hobbyist who does not know the correct breeding methods for bearded dragons would not be able to read the animal's behaviour, such as signs to lay.
- Bearded dragons should be allowed to breed, as it is part of their natural behaviour. The Animal Welfare Act 2006 states that animals must be able to exhibit their natural behaviours.

42 Understanding command words

1 *Example of a complete answer to question 4:*
It is fit for purpose due to the water-resistant material it is made of. It has the potential to be reused, making it good value and reducing its impact on the environment… The material is easy to clean which will help with good hygiene and maintaining animal health; water is easy to top up and replace. It is made out of plastic and metal, so could break if dropped. The ball-bearing that controls the drip function can become stuck, which would prevent the water bottle from functioning correctly.
Example of a complete answer to question 5:
Water bottles for animals are usually made out of plastic, while water dishes can be made of plastic, metal or ceramic. Both bottles and dishes come in a variety of sizes, making them suitable for a range of animals. The method selected will depend on the animal it is for. Small mammals such as rodents, rabbits and guinea pigs naturally drink from ground level, so a dish would serve this purpose well … However, water dishes can be hard to keep clean due to the digging behaviours of some of these animals, making a water bottle a much better choice. The measurements on the side of water bottles and the clearly visible water level provide an easy way to monitor how much water intake your animal is getting. Another method of providing water is a spigot watering system, which is made of metal and provides a constant supply of water. This is generally used for livestock mammals, as it easy to use and does not need changing daily, which works well on a busy farm.

2 *Example answer:*
Gene manipulation techniques could be used by our society to develop disease-resistant animals in two ways. First, genetically engineered animals would provide valuable models with which to investigate disease progression and to evaluate this approach to controlling the disease. Second, these animals could be introduced into breeding programmes for livestock to aid the food production industry. However, there are concerns regarding gene manipulation techniques; the main one, of course, being the welfare of the selected animals. There is also the potential for diseases to spread, especially those that are zoonotic (can be transmitted from animals to humans). If successful, gene manipulation might not only affect the animal industry (for example, veterinary interventions not being required so much and vaccinations not being needed), but also human health, as it could result in the discovery or development of new treatments.

43 Short-answer questions

Example answers (only two required):
- Vocalisation
- Scent marking
- Flight response
- Hopping
- Circling
- Leg thumping
- Chewing.

44 Longer answer questions

Example answer:

In his study of the genetics of pea plants, Gregor Mendel performed dihybrid crosses, which are crosses between organisms that differ in two observed traits. He discovered that the combinations of traits in the offspring of his crosses did not match the combinations of traits in the parent organisms. From this study he formulated the Principle of Independent Assortment, which is that different genes independently separate from one another when reproductive cells develop.

Independent assortment occurs due to meiosis, which is the type of cell division that reduces the number of chromosomes in parent cells by half to produce four reproductive cells, called gametes.

45 Essay-style questions

Your plan might include the following points:

There are many concerns regarding use of genetic modification in breeding of animals, including:
- **Animal rights** – animals are manipulated for human gain.
- **Public safety** – dangers that the animals being used could pose to humans, especially disease transmission.
- **Animal welfare** –
  - Pros: it could benefit animals by improving resistance to disease and removing breed characteristics that could cause health problems for the animals.
  - Cons: breeds adapted to increase food production, but causing harm or discomfort to animals, e.g. pigs bred to grow faster causing issues for their health.

Unit 2: Animal Biology

46 Cellular ultrastructure

1. in the cell nucleus
2. produce ATP through respiration

47 The fluid mosaic model

Separate the contents of the mitochondria from the cytoplasm. Control the movement of molecules in and out of the mitochondria.

48 Microscopy

15/0.0075 = 2000

49 Cellular control

32 pairs of chromosomes

50 Cell transport

diffusion

51 Active transport

ATP

52 Animal tissue types

epithelial, connective, nervous, muscle

53 Epithelial tissue

Simple squamous; found in the alveoli

54 Connective tissue

Any two from:
- bone
- cartilage
- blood

55 Structure and function of muscle tissue

1. (a) skeletal
   (b) cardiac
   (c) smooth
2. *Any two from*
- reproductive tracts
- the blood vascular system
- bladder
- bowels

56 Structure of nervous tissue

1. sensory, motor, interneurons
2. Myelin sheath improves speed of transmission within neurons; long neurons need myelin to ensure rapid transmission of impulse; interneurons are short; and do not need insulation

57 Fast and slow twitch muscle fibres

1. *Any three from:* fast twitch fibres contract rapidly/quickly; are lighter in colour than slow twitch fibres; have a lower density of capillaries; do not rely on oxygen; use anaerobic respiration; have a low density of mitochondria.
2. via the blood vessels

58 Muscle contraction

There would be less muscle contraction. This is because there would be fewer myosin binding sites unblocked if there was insufficient calcium available.

59 Functions of the skeleton

A cranium and vertebrae.

60 Classification, structure and function of bones

1 carpal 2 patella/knee cap 3 vertebra

61 Joints and muscles

hinge, pivot, ball and socket, gliding, condyloid, saddle

62 The integumentary system

The fat in the subcutaneous layer will provide insulation while the dermis contains blood vessels which can constrict to reduce blood flow to the surface of the skin, thus reducing heat loss. The fox will also have a thick undercoat which will provide further insulation.

63 Structure and function of feather types

contour, down and semiplume feathers.

64 Methods of locomotion

Any two from:
- long limbs
- lightweight skeleton
- no collar bone to allow for more movement.

65 Musculoskeletal adaptations and disorders

1. Possible causes include injury, incorrect weight and insufficient nutrition.
2. Will cause stiffness, difficulty rising and uneven gait. This could lead to reduced mobility, obesity and therefore stress on the joints.

66 Nutrients and digestion

Any four from: proteins, fats, fibre, water, vitamins, minerals, carbohydrates

67 Specialised digestive systems

1 fibre
2

Part of stomach	Function
Omasum	**Some water and salts are absorbed**
Reticulum	Catches foreign objects and works alongside the rumen
Rumen	**Microbes secrete enzymes such as cellulase to break down cellulose. Saliva helps to ensure there is a suitable environment for micro-flora (bacteria) to survive**
Abomasum	Where enzymes break up food – the 'true stomach'

68 Oral cavity adaptations

large canines for ripping meat into chunks

69 Regulation of blood glucose

Increased blood glucose is detected by the pancreas which releases insulin from the beta cells in response. The insulin encourages the conversion of glucose to glycogen which is then stored in the liver, reducing the amount of glucose in the blood.

70 Digestive disorders

Ingesting tools can lead to punctures or obstructions in the digestive tract wall. If undiagnosed, this can lead to illness and death.

71 Action potentials

1 If the threshold is reached, an action potential is triggered. This opens more Na+ channels and depolarisation occurs. This is where the cell becomes more positively charged on the inside than out.
2 At the peak voltage of around +40 mV, Na+ channels close and voltage-gated K+ channels open. K+ ions diffuse out, causing repolarisation of the cell.
3 This causes hyperpolarisation of the neuron. More K+ ions are on the outside than Na+ ions are inside.
4 A refractory period allows the sodium- potassium pump to return K+ ions to the inside and Na+ ions to the outside, returning the neuron to its normal polarised state.

72 The nervous system

Movement of the whisker causes movement of the free nerve ending which is attached. As the free nerve ending moves, ion channels open, initiating an action potential. This travels along the afferent neuron, then across the synapse to the relay neuron in the CNS, and along the spinal cords to the brain.

73 The autonomic nervous system

1 acetylcholine
2 sympathetic division

74 Receptors and sense organs

the skin

75 The structure of the eye

The cornea does not have a blood supply so needs to get its nutrients from elsewhere. The aqueous humour is liquid so it allows nutrients to diffuse across it easily.

76 How the eye works

1 on the pupil
2 opsin and retinal

77 Eye adaptations

1 binocular vision
2 Advantages: Better depth perception and no blind area directly in front.
Disadvantages: smaller visual field and need to move head more to review surroundings.

78 Common neurological disorders

The animal may be nervous, restless, shaking or salivating.

79 Structure and function of blood

Help your body repair by stopping bleeding after illness or injury

80 The circulatory system

1 oxygenated blood away from the heart
2 The aorta has a thicker and more muscular wall than the vena cava. The lumen of the vena cava is larger than the lumen of the aorta. The vena cava has valves, while the aorta does not.

81 The cardiac cycle

1 the aorta
2 the sinoatrial node

82 The respiratory system

in the alveoli

83 Respiration

lactic acid build up due to anaerobic respiration

84 Gas exchange and blood proteins

1 They are very thin with a large surface area.
2 oxygen and carbon dioxide

85 The lymphatic system

1 in the thymus
2 B-cells

86 Circulatory system disorders

To ensure affected dogs are not entered onto the breed pool and not passed on.

87 Male reproductive system in mammals

Where the sperm matures and is stored

88 Female reproductive system in mammals

1 (a) provides protection for the reproductive system from infection
(b) where the fertilised egg implants; it stretches to accommodate growth of the foetus, and later contracts to expel foetus at birth
2 bicornuate

89 Fertilisation and gestation length in mammals

114 days

90 Reproductive hormones

1 spermatogenesis
2 oestrogen

91 The oestrous cycle

During pro-oestrous the level of oestrogen increases. Once oestrous begins, the level of oestrogen decreases.

92 Roles of hormones in parturition and lactation

1 oxytocin
2 prolactin triggers mammary glands – lactation; oxytocin releases milk and is stimulated by offspring suckling

93 Reproductive system in birds

1 one ovary
2 infundibulum, magnum, isthmus, uterus, cloaca

94 Structure of an egg

Example answer:
The shell provides a hard outer layer which protects the embryo from chemical toxins. The amnion is a shock absorbing fluid which protects the embryo from physical damage.

95 Egg formation

in the infundibulum

96 Embryo development

1 21 days
2 *Any four from:* availability of light (photoperiod), food availability, diet, type of breed, housing, husbandry, moulting, age and weather

97 The excretory system

1 The urinary system removes waste products.
2 the bladder

98 Nephrons

1 renal (Bowman's) capsule
2 loop of Henle, distal convoluted tubule
3 proximal convoluted tubule, loop of Henle

99 Osmoregulation

1 pituitary gland
2 water content in blood drops; detected by osmoreceptors in hypothalamus; pituitary gland releases more ADH into the blood; permeability of collecting duct walls increases; more water is absorbed into the blood via osmosis; less water lost in urine.

100 Nitrogenous waste removal

1 Urea
2 kidneys

101 The thermoregulatory system

1 control of body temperature
2 negative feedback

102 Warming and cooling mechanisms

Arterioles constrict, so less blood flows to the skin which decreases heat loss.

103 Adaptation and variation

behavioural

104 Modern technology and classification

1 DNA and RNA
2 DNA hybridisation

105 Living organism classification

1 Animalia and Prokaryotae
2 Kingdom, Phylum, Class, Order, Family, Genus, Species

106 Classification difficulties

Armadillos have an ossified outer armour, but a complete endoskeleton. The presence of a partial exoskeleton is similar to the skeletons of chelonia.

107 Your Unit 2 Exam

Example answer:
Allow about 1 minute per mark, to leave a little time for checking your answers.
(a) So allow about 2 minutes for a 2-mark question
(b) 8 minutes for an 8-mark question.
   (In practice, you might be able to answer short answer questions much more quickly, and use the time saved to plan the structure of your longer answers.)

108 Short-answer questions

1 rumen, omasum, abomasum and reticulum
2 abomasum

109 'Complete' and 'define' questions

Fibrous	Bones are held together by fibrous connective tissue which does not allow movement between them.
Synovial	**These are freely moveable and occur where two bones meet.**

110 Data questions

$$\frac{75 + 78 + 82 + 90 + 88}{5} = \textbf{82.6 beats per minute}$$

111 'Explain' questions

Your answer could include any of the following:
Naturally occurring variations in animal's DNA leading to some of the animals inhabiting a different niche; loss of habitat leading to changes in food availability leading to evolution of feeding behaviours; climate change impacting on breeding patterns and courtship rituals.

112 'Describe' questions

Example answer:
Haemoglobin's main function is to transport oxygen around the body. The oxygen is transported from the lungs to the body's tissues and then carbon dioxide is transported out of the tissues back to the lungs.

113 'Discuss' questions

Example answer:
If body temperature is too high, the blood vessels supplying capillaries in the skin dilate and become wider. This results in more blood flowing through the capillaries near the body's surface so that heat is lost. If body temperature is too low, the blood vessels supplying capillaries in the skin constrict and become narrower. This results in less blood flowing near the surface so less heat is lost.

114 'Compare' questions

Your answer should include some or all of the following content and should compare and contrast the features you mention:
Parasympathetic: decreases overall physical activity.
Decreases heart rate, ventilation rate and increases digestion.
Uses acetylcholine.
Rest and digest.
Sympathetic: prepares the body for physical activity.
Increases heart rate, ventilation rate and decreases digestion.
Uses noradrenaline.
Fight or flight.

Unit 3: Animal Welfare and Ethics

115 Animal welfare

Example answer:
Education – if owners are made aware of the specific husbandry requirements (e.g. exercise requirements) there is less chance of the animal suffering.

116 Animal housing

Example answer:
The accommodation is suitable for two rabbits because of its size and design. The enclosure comprises an outside and an indoor area, and there is plenty of space. It has access to natural areas (grass), and provides protection from the elements in the top area, as well as a warm bedding area. There is plenty of ventilation and lighting. It looks accessible for cleaning, feeding, etc.

Disadvantages include the materials the enclosure is made from: the wire will not provide protection from larger animals such as foxes or dogs. The wood can be gnawed by the rabbits and sharp splinters may injure them. The wood could deteriorate due to absorbing rainwater and urine. The accommodation does not look strong or long-lasting. The rabbits could escape by digging, or predators could dig to enter.

117 Suitable environment

Example answers:
Substrates: Provide shavings (suitable throughout the year for ferrets), as these are absorbent, which will ensure urine and rain do not penetrate the hutch, and will prevent water pockets developing. Place additional straw in the bedding area to ensure the ferrets stay warm. Animal blankets/fleeces can also be added.
Heat sources: There is no need for heat lamps or other heat sources for this species as they are adaptable to the cooler environments and these could pose safety risks if used.

118 Balanced diets

Example answer:
A balanced diet that is appropriate to the particular species helps maintain the overall health of an animal. It ensures the animal has enough energy to complete its normal daily activities, e.g. exercise, and that it grows and develops well.

119 Nutritional problems

Example answer:
1   Metabolic bone disease (MBD) is a potential problem caused by deficiency in vitamin D_3 and/or calcium and phosphorous.
2   The correct environment and diet must be provided, e.g. UVB lighting, correct diet and supplementation of calcium and phosphorus.

120 Dietary needs of ruminants

Example answer:
Ruminant animals are a type of herbivore. Species include *two of*; sheep, goats and deer. These have a different digestive system to non-ruminant animals. The stomach in a ruminant animal has four compartments, and the animals need a long time to graze and digest their food.

121 Dietary needs of hindgut fermenters

Example answers:
1   A caecotroph is a special type of faeces, expelled from the anus as a small pellet.
2   It is different to usual faecal matter, both in consistency and due to the key fact that some animals (e.g. rabbits) need to re-ingest it to ensure they acquire the correct nutritional values from their feed.

122 Normal behaviour patterns

Answers will vary according to the animal chosen. See the example behaviours listed on page 122, or think about behaviours that the animal would display in a wild environment, e.g. hunting, foraging, playing, etc.

123 Behaviour and feeding

Example answer:
Herbivores eat plant material, so they may need to spend a long time grazing or browsing to ensure they obtain the correct amount of nutrients. Other herbivorous species may forage for seeds or root for plant material, such as fungi and roots underground.

124 Circadian rhythms

Example answer:
A nocturnal animal is active at night and sleeps during the day. Examples include: hedgehog, badger, some owl species.

125 Environmental enrichment

1   *Your answer should state that environmental enrichment improves animal welfare. It should give examples of the different elements that contribute to high levels of welfare, such as: mental stimulation, exercise, a natural way of feeding, keeping the animal occupied, creating a natural environment, encouraging natural and social behaviours, providing sensory experiences, reducing the likelihood of stereotypic (abnormal) behaviours.*
2   *Example answer:* Dogs enjoy using feeding devices that they have to manipulate and play with before they get to their food, e.g. treat-dispensing toys and puzzles. You can hide food around a hamster's cage to encourage the hamster to search for it.

126 Exercise

Example answer:
Exercise reduces the likelihood of diseases such as diabetes and heart conditions. It helps the animal to live longer. It also helps stimulate the animal's mind, therefore reducing the likelihood of abnormal behaviours.

127 Solitary animals

Example answer:
Syrian hamsters are solitary animals, so they should be housed alone. They can be housed together for mating or when young hamsters are not fully weaned, but they should be separated as soon as possible after weaning.

128 Presocial animals

Example answer:
Elephants are social animals so they should be housed in a group. They can be separated for short periods of time for activities such as health checking but, during separation, interactions should be allowed, for example, in the form of seeing one another through bars.

129 Presocial and eusocial animals

Example answer:
Eusocial animals meet all three of the criteria for social behaviour (communal living; taking care of others' young; sharing the labour, e.g. protecting others in the group and gathering food), whereas presocial animals can meet any of these three criteria, but not all at the same time.

130 Preventative care and treatment

1   Endoparasites live inside the host, e.g. tapeworms – 'endo' means within.
    Ectoparasites live outside the host, e.g. fleas and ticks – 'ecto' means outside.
2   Good husbandry management will help prevent and control the spread of parasites, e.g. providing clean water, cleaning of water and feed bowls, as well as any equipment, enrichment devices and the animal's enclosure itself. Provide good nutrition to all animals. Animal workers need to undertake good hygiene, e.g. using PPE, and always washing hands before and after handling animals.

Example endoparasite: stomach worm (tapeworm)
(i) Cattle: calves can be wormed when 'turning out'; provide clean grazing for cattle.
(ii) Sheep: as with cattle, ensure sheep are put out on pasture with low or no worm infestation
(iii) Rabbits can also be administered dewormers.
Example ectoparasite: flea
(i) Cattle: ectoparasiticides are available for the control and treatment of ectoparasites such as fleas, ticks and mites. These are external treatments and come in different forms: 'dipping', dust powders, ear tags, pour-ons and sprays.
(ii) Sheep: similar treatments are available as for cattle, but ear tags are unlikely to be used with sheep.
(iii) Rabbits: if rabbits have an infestation of fleas, flea combs can be used, as well as a rabbit-friendly flea powder. Regular grooming can help in prevention and monitoring.

131 What are ethics?

Example answers:
1 Personal feelings are those held by an individual, regardless of whether their beliefs are the same as the rest of society. For example, a vegetarian might believe that animals raised for food experience poor welfare, even if the correct laws are being followed.
2 Legislation protects animals in all settings against poor animal welfare practices, e.g. improper care. Animals found in zoos, farms and laboratories, and pets are all covered.

132 Ethical theories

Example answers:
Utilitarianism: may be in favour of animal testing of medicines (for humans or other animals) and recommend that sacrificing a few animals to increase the survival of a large number of animals is the correct moral course to take.
Sentience-centred approach: could be used to justify the use of invertebrate animals, such as fruit flies and other insects, in research because they do not feel pain in the same way as vertebrate species. The approach might condemn using vertebrate animals such as rodents, dogs or primates, which have been proven to feel pain.
Relational importance: may be happier to use animals that humans don't usually have relationships with (such as insects and reptiles) rather than those species of mammals and birds that we have domesticated.

133 Exploration of ethical frameworks

Example criteria for the matrix:

Respect for ...	Health and welfare	Freedom and choice	Fairness
Consumers	**Quality produce and food safety** Pigs that are bred with high welfare standards are likely to produce better meat.	**Options based on information supplied** Good farmers have a good reputation, and consumers may opt for meat from these farmers.	**Food availability** **Food choice** More food choice for consumers.
Farmers	**Good income and working conditions** Provides work for pig farmers and their employees.	**Freedom of action** Farmers can choose to work in this industry.	**Compliance with legislation** All farmers must follow welfare legislation.
Farm animals	**Animal welfare and conservation** Pigs are in an abundant supply in the UK (not endangered). They are social animals so should be housed in groups.	**Behavioural freedom** Pigs do not choose to be kept in the conditions they are housed in, nor to breed with the pigs that are chosen for them to breed with.	**Intrinsic value of the animals** Pigs are sentient beings, so should not be bred for the meat industry.

Example discussion based on the matrix:
Farmers are likely to continue breeding pigs for use in the food industry. The industry provides many people with jobs, and provides many consumers with different food options. Pigs are in plentiful supply and are not endangered. From a fairness point of view, pigs are intelligent animals and can experience emotions, but farmers must ensure that their welfare needs are met, e.g. by allowing them to live in a social grouping, which is also adhering to legislation.
Note that there is no one fixed answer that could come out of examining the matrix. It is a method of making sure that you consider how the issue affects a range of different stakeholders from different ethical viewpoints.

134 Using ethical frameworks

Example answer: In a veterinary practice it is commonly regarded as ethical to euthanaise a seriously/terminally ill animal to prevent further suffering.
You don't need to give a full answer for this revision question, but a full answer would include:
• the impact of each possible course of action on the owner, the cat and the nurse
• reference to a number of ethical theories when looking at the issues involved in both the actions and the potential outcomes
• consideration of how each issue is related to the others
• links between the different aspects of the scenario, supported by evidence from the case study and outside, e.g. codes of practice
• judgement on the best course of action, which has been clearly reached by drawing together the reasoning that you have made.
An ethical framework focuses on what is perceived as right or wrong in a given situation. It can also look at the standards within a specific field to help determine whether something is ethical or unethical.

135 Consumer ethics

Example answer:
Anthropomorphism means interpreting an animal's behaviour as if it were a human. The dog may appear to be smiling, but its bared teeth are more likely to be a warning. Someone who misinterprets this behaviour may not allow the dog sufficient space to feel comfortable, and the dog may become aggressive and perhaps bite.

136 Definitions of animal welfare

Your answer might include the following:
1 The five welfare needs should be met to a good standard, ducks are healthy and showing no signs of injury, illness or distress, sufficient space according to RSPCA requirements, suitable socialisation, protection from predators, natural environment.
2 Sufficient feed and resources to ensure high productivity, easy to maintain and clean, cost effective.
3 Appropriate food, water and exercise, suitable preventative treatments, good disease control and prevention, low stress environment.

137 Animal welfare legislation

Your answer should focus on providing for and not neglecting the guinea pig's five welfare needs. Example answer:
Guinea pigs must be provided with a sheltered area that is of sufficient size to allow the animals to move around and exercise. The housing should be safe and secure and not let prey animals enter, and it should provide a hiding area. Guinea pigs should be housed in groups, as they are a social species. Enrichment should be provided to keep them active and allow for normal behaviours, such as grazing. A correct and complete diet should be provided of hay, fresh grass and fresh vegetables. They need to have Vitamin C in their diet, as they cannot make it themselves.

138 Role of the government in animal welfare

Example answers:
1 The local authority will process the application for the new pet shop. They will check that the shop meets the requirements of the legislation before issuing a licence.
2 They will carry out checks on the welfare and safety of the animals and workers, for example:
   • correct species and numbers of animals being kept
   • good levels of welfare are maintained: living conditions, diet, breeding, etc.
   • correct facilities provided, e.g. exercise areas
   • health and safety of employees and visitors to the shop.

139 Legislation and codes of practice

Example answer:
Professor F.W. Rogers Brambell investigated the welfare of farm livestock and published his report in 1965, in which he made many recommendations for the welfare of livestock animals, for example, that they should have enough space to 'stand up and turn around'. The conditions that he recommended were the basis for the Farm Animal Welfare Council's five 'ideal state' of animal welfare, known as the 'Five Freedoms'. These have contributed to more recent animal welfare legislation, particularly the Animal Welfare Act 2006.

140 Pet animals

Example answer:
Licence being revoked, a ban on keeping animals or a pet shop, fines, community service or unpaid work, or imprisonment, depending on the severity of the offences.

141 Dogs

(a) not before one year old
(b) six
(c) 8 weeks

142 Riding and boarding establishments

1 Riding Establishments Act 1964 and 1970
2 *Example answer:*
   The applicant is over 18 and not disqualified from running a riding establishment, owning a pet shop or having custody of animals; they are suitable and qualified; the horses are in good health to be used for riding and teaching; there are fire evacuation procedures in place; housing is suitable; suitable insurance is in place; the establishment keeps the required records.

143 Farmed animals

Example answer:
It is specific to animals at market and ensures their welfare with regards to their conditions, food and water regardless of their age. It aims to ensure that no animals experience harm, injury or suffering. The local authority can inspect animals at markets at any time.

144 Transporting animals

Your answer should include the main elements of the transport legislation as well as the competences required of the farmer and/or transporter. For example: ensuring sufficient cleanliness in the lorry – you would expect the correct bedding and amount to be supplied, with regular removal of urine and faecal matter (as a minimum before/after transportation, depending on the length of the journey).

145 Wild animals in zoos and private collections

Any four animals listed in the schedule of dangerous wild animals found listed in the Dangerous Wild Animals Act 1976, e.g.: red panda, gorilla, giant armadillo, caiman, cobra, wandering spider, gibbon, hyena, tapir.
For a full list of the species, refer to the legislation.

146 The fur trade

Example answer:
Mink fur can be obtained either from a fur farm or wild caught, as they are not classed as endangered in the wild by the IUCN Red List. They are a species that is commonly captive-bred for fur farms.

147 Animals in science and education

Example answer:
The stakeholders include the staff and students at the college, and the dog. The staff and students would benefit from being able to demonstrate and witness appropriate techniques for obtaining a variety of data, indicating animal health. The dog may experience stress from being in a college setting, surrounded by a large group of people, and possibly from having the techniques practised on it by many students.

148 Animals used for entertainment

Your answer should refer to meeting the five animal needs.
Example answer:
Accommodation must be suitable for the goat when not being trained. It must contain substrates, e.g. shavings as these are absorbent. Additional straw should be placed in the bedding area to ensure warmth. The goat must have access to shelter as well as a secure outdoor area. Proper food, water and veterinary care must be available when required. The goat should have sufficient rest breaks and, as goats are social animals, it should be housed with at least one other of its species,

149 Wildlife and conservation

Example answer:
Yes, under the Wild Mammals (Protection) Act 1996, it is legal to kill a seriously injured wild animal to relieve its suffering, as long as it is done in a humane way.

150 Reasons for and methods of killing animals

Example answer:
It renders the animal unconscious, meaning they have no awareness, feeling or thought on the event, ensuring the death is as stress- and pain-free as possible.

151 The role of the animal welfare inspector

Your answer should include five duties from the spider diagram on page 151, or any other appropriate duties from your studies or your own experience. Key words you might include are:
- respond
- assess
- educate
- prepare case file
- administration.

152 Welfare appraisals

Your appraisal should cover all five animal welfare needs, and address all the questions on the form.

153 Measuring behaviour

1 *Example answer:*
    Pacing or circling – could mean the animal's enclosure is too small or there is not enough enrichment provided.
    Excessive vocalisation – e.g. dog whimpering, could indicate stress or injury.
    Other behaviours could include: bar biting, crib biting, feather plucking, glass surfing.
2 An ethogram is a list of behaviours, and their descriptions, that a particular species will display.

154 Welfare appraisal measuring techniques

Example answer:
Excessive urine/faeces in the enclosure could indicate the lack of husbandry management and general care of the animal, or untreated disease.
A low body score/weight could indicate incorrect diet being fed, or untreated disease.
You might also have described any of the other indicators in the spider diagram on page 154.

155 Welfare appraisal action plans

Example answer:
The inspector might decide to reduce the number of dogs in the owner's care, reasoning that if the owner had just one or two dogs, they would find caring for them more manageable. Further welfare inspections will be completed in the near future to review whether the dogs' welfare has improved.

156 Your Unit 3 set task

- The set task brief, and set task information for your research to prepare for Part B; about 6 hours
- 4; 3 hours
- four sides of A4 from Part A, stimulus material booklet

157 Reading a brief

1 *The key points to highlight are shown below:*

> **Task brief**
> You are required to carry out research into the scenario provided in the task information below. You should consider the following areas in relation to the scenario:
>
> - legislation and regulations relating to the animal species
>
> - policies and practices relating to the setting and linkages to the welfare requirements of that species.
>
> **Task information**
> Ridgemoor Boarding Kennels in Westshire has recently been inspected by the local authority and found to be unsuitable. Ridgemoor Boarding Kennels has appointed a new manager to review and improve the situation.
>
> You have been asked by the new manager to assist in the appraisal of Ridgemoor Boarding Kennels.

2 *Example answer:*
Areas to research:
- Legislation and regulations relating to kennels and dogs:
    - Animal Welfare Act 2006
    - Animal Boarding Establishments Act 1963
    - Dangerous Dogs Act 1991
    - Breeding of Dogs Act 1973 and 1991
    - Breeding and Sale of Dogs (Welfare) Act 1999
    - Clean Neighbourhoods and Environment Act 2005 (regarding walking of the dogs)
    - The Microchipping of Dogs (England) Regulations 2015
    - The Control of Dogs Order 1992
- Policies and practices in a kennels – procedures to cover:
    - Diet and feeding
    - Cleaning of kennels
    - Maintenance of kennels
    - Health monitoring of dogs
    - Administering medication
    - Human contact
    - Exercise
    - Temperature and bedding
    - Supervision and reporting
    - Record keeping
    - Admittance procedures, including vaccinations check
    - Treatment of dogs that become ill during stay
    - Procedures for disease control
    - Fire and emergency evacuation procedures
- Stakeholders: dogs, owners, kennel owners, kennel staff, local authority, animal welfare organisations, general public

158 Carrying out research

Example answer:
The Animal Boarding Establishments Act 1963
Applies to businesses where animals are boarded for the purpose of making profit. Requires a licence from the local authority. Boarding establishments defined as 'those premises, including private dwellings, where the business consists of providing accommodation for other people's cats and dogs'. Before issuing a licence, the local authority must consider the suitability of the conditions present at the boarding establishment, including:
- Accommodation must be suitable in relation to:
    - Construction – must protect from the weather, nontoxic, escape-proof, free from hazards, floor non-slip.
    - Size – large enough to allow separate sleeping and activity area; large enough to allow dog to turn comfortably and wag tail without touching sides.

- Number of occupants – allow dogs from the same home to be kept together as long as no behaviour issues.
- Exercising facilities – must allow for exercise to be provided away from the kennels.
- Temperature – must avoid extreme temperatures. There should be an area that is above 10°C and below 26°C.
- Lighting – should have access to indoor and outdoor areas and allow natural light cycle to happen. If not possible, indoor lighting must match this.
- Ventilation – indoor areas must be provided with sufficient fresh air.
- Cleanliness – all areas dogs have access to should be kept clean and disinfected regularly.
- Adequate supply of suitable food, drink and bedding material:
  - Must have access to water at all times.
  - Must be provided with an adequate diet that fulfills their nutritional needs.
  - Should be fed at least once a day, ideally twice.
  - Bedding must be provided.
- Dogs must be adequately exercised and visited at suitable intervals.
  - 30 minutes exercise must be provided away from the kennels.
- Precautions must be taken to prevent and control the spread of infectious and contagious diseases, including isolation facilities.
- Records must be kept of any animal that comes into the establishment, and must be available for inspection at all times.

159 Making notes

Animal Boarding Establishments Act 1963

Requires anyone who wishes to keep a boarding establishment to be licensed by a local authority and to abide by the conditions of the licence.

The local authority will consider whether the establishment is able to ensure that the following provisions are met:
- Accommodation must be suitable in relation to: construction, size, number of occupants, exercising facilities, temperature, lighting, ventilation and cleanliness.
- There must be an adequate supply of suitable food, drink and bedding material.
- Animals must be adequately exercised and visited at suitable intervals.
- Precautions must be taken to prevent and control the spread of infectious or contagious diseases, including isolation facilities.
- Records must be kept of any animals that come into the establishment, and these must be available for inspection at all times.

160 Responding to stimulus material

Your own response based on the stimulus material you chose.

161 Writing an appraisal report

Example answer:

With regard to feeding, Ridgemoor Boarding Kennels are correctly obeying the Animal Welfare Act 2006 and the Animal Boarding Establishments Act 1963 by providing an adequate supply of food and water, as evidenced in their daily staff procedures (use of whiteboard to record feed given).

162 Evaluating information supplied

Examples of information you would need to see in order to carry out a comprehensive appraisal:
- Boarding Establisment Licence – either a paper copy or seen in person – it should be on display at the premises.
- Inspection report of the accommodation, ideally done in person, so you can thoroughly check cleanliness across all areas. This would be hard to ascertain from photos.
- Feeding plan for current occupants – either paper copies, photos or seen in person. These will need to be checked against the records for individual dogs.

- Exercise facilities and records – either photos or ideally in person to allow for a more in-depth check. An actual inspection will allow you to see dogs using the facilities or being walked, providing a true indication of the dogs' welfare and whether the legislation is being obeyed.
- Control measures plan for preventing spread of infectious and contagious diseases, including isolation facilities. Can be seen in person or on paper. Ideally, the facilities need to be inspected in person to check the control measure meet the required standards.
- Animal/owner records. Paper copies would be adequate to meet legislation, but of course inspecting these in person allows you to check filing systems and that records are being used effectively.

163 Producing an action plan

Example answer:

Issue to address	Fix broken drain covers in Kennels 3 and 14.
Action	The drain covers need to be replaced by the correct type – product no. 3721.
How	The covers can be easily sourced from the local hardware store.
Why	Maintenance staff to source and fit the covers.
Timescale/ urgency	These kennels not to be used until fixed. Need to source new covers and replace within one week (by 4th March).

Reason for action:

These kennels cannot be used until covers are fixed as they may cause injury to dogs or staff. Because of the financial impact on the business (no income from these two kennels while out of use), ideally they need to be fixed as soon as possible. One week is ample time to achieve this.

164 Exploring ethical issues

Your list might include:
- Large number of people and animals in one place – heat and noise issues, spread of diseases, dogs fighting.
- Breed issues associated with German shepherd dogs , e.g. hip dysplasia.
- Should we even own pet/dogs, let alone show them for personal gain?
- Benefits of showing dogs: improving breed standards, aiding pet industry, tradition.
- Social element of showing: meeting both animal and human social needs.
- Benefits for Ridgemoor Boarding Kennels of showing their dogs, e.g. financial gains, reputation.
- Stress levels of the dogs who are not used to being shown (this will be the first time).

Notes

Notes

Notes

Notes

Published by Pearson Education Limited, 80 Strand, London, WC2R 0RL.

www.pearsonschoolsandfecolleges.co.uk

Copies of official specifications for all Pearson qualifications may be found on the website:
qualifications.pearson.com

Text and illustrations © Pearson Education Ltd 2017
Typeset and illustrated by Kamae Design
Produced by Out of House Publishing
Cover illustration by Miriam Sturdee

Picture research by Alison Prior

The rights of Natalie Betts, Laura Johnston and Leila Oates to be identified as authors of this work have been
asserted by them in accordance with the Copyright, Designs and Patents Act 1988.

First published 2017

2024

15

British Library Cataloguing in Publication Data
A catalogue record for this book is available from the British Library

ISBN 9781292150000

Acknowledgements
The author and publisher would like to thank the following individuals and organisations for permission to
reproduce the following:

Text
Page 5 from "Role of microRNAs in skeletal muscle development and rhabdomyosarcoma (Review)" by
Huiming Ju; Yuefei Yang; Anzhi Sheng; Xing Jiang in Molecular Medicine Reports. Published by Spandidos
Publications © 2015; pages 87 & 88 from the Merck Veterinary Manual, Online version, Scott line ed. copyright
© 2017 by Merck & Co. Inc., Kenilworth, NJ. All rights reserved. Used with permission. Available at http://www.
merckvetmanual.com Accessed May 31, 2017; page 139 from The Farm Animal Welfare Council (FAWC) 1979.
Crown copyright under the HYPERLINK "http://www.nationalarchives.gov.uk/doc/open-government-licence/
version/3/" Open Government Licence. (http://webarchive.nationalarchives.gov.uk/20121010012427/http://www.
fawc.org.uk/freedoms.htm).

Photographs
(Key: b-bottom; c-centre; l-left; r-right; t-top)123RF.com: 17cr, 24, 36, 110, imagex 64, marosbauer. 57r, prensis
17l, silense 15, verastuchelova 127b; Alamy Stock Photo: 360b 135c, Arco Images GmbH 20b, 30l, Avico Ltd
148, Blickwinkel 155, Dirk Erken 145, FLPA 5r, 16, 144, G Dobner CO 115, GROSSEMY VANESSA 20t,
Image Source 136bl, imageBROKER 116, John Eccles 141, Manfred Grebler 142, Monkey Business 129l,
Papilo 149, Peter Titmuss 151, REUTERS 37, Shaun Higson colour 136tl, Simon Belcher 96, Simon Belchers
27, Stockbroker 130t; Compassion in World Farming: 135b; Getty Images: emholk 10, Margo 22l, 22r; Mary
McCartney/PETA: 135t; Nature Picture Library: Yves Lanceau 5l; Pearson Education Ltd: Jon Barlow 11, Lord
and Leverett 74tr; Shutterstock.com: 2630ben 106, 41 79 (a), 528703 74tl, Africa Studio 8, 147, Alan Jeffery
129r, Alexander Mozymov 14t, Angel Soler Gollonet 54t, Atthapol Saita 127t, Bildagentur Zoonar GmbH
130b, bluedog studio 7c, BMJ 77l, Budimir Jevtic 70, CathyKeifer 74bl, CatMicroStock 94, Choksawatdikorn
54b, Chris Fourie 122, Christopher Meade 54cl, Claudio Divizia 48, David Litman 54c, Eric Isselee 7l, 7r, ESB
Professional 156, FAUP 128, Filed Image 143, gorillaimages 26b, Guy J. Sagi. 62, inkwelldodo 125t, Jamie Hall
17r, Maggie Harrison 22b, Kateryna Kon 54tc, Lee Torens 77tr, Maggie 1 12, Marko Sobotta 14b, MENATU 1t,
Michelle Lalancette 57l, Mikhail Pogosov 1b, Mikkel Bigandt 86, Monika Wisniewska 26t, 78, mountainpix 123,
Natali Glado 64b, Olga Kot Photos 146, Pablod Debat 89, pleny m 79 (d), PrakapenkaAlena 126, RedVila 136tr,
rickyd 17b, Rob van Esch 100, Rosa Jay 77t, royaltystockphoto.com 79 (c), Soultkd 140, THANISORN PANYA
121b, toeytoey 79 (b), unol 23, VitStudio 6, Weldon Schloneger 121t, WilleeCole Photography 30r, Yi-Chen
Chiang 76, Zoltan Major 126bAll other images © Pearson Education

Notes from the publisher
1. While the publishers have made every attempt to ensure that advice on the qualification and its assessment is
accurate, the official specification and associated assessment guidance materials are the only authoritative source
of information and should always be referred to for definitive guidance. Pearson examiners have not contributed
to any sections in this resource relevant to examination papers for which they have responsibility.

2. Pearson has robust editorial processes, including answer and fact checks, to ensure the accuracy of the content
in this publication, and every effort is made to ensure this publication is free of errors. We are, however, only
human, and occasionally errors do occur. Pearson is not liable for any misunderstandings that arise as a result of
errors in this publication, but it is our priority to ensure that the content is accurate. If you spot an error, please
do contact us at resourcescorrections@pearson.com so we can make sure it is corrected.

Websites
Pearson Education Limited is not responsible for the content of any external internet sites. It is essential for
tutors to preview each website before using it in class so as to ensure that the URL is still accurate, relevant and
appropriate. We suggest that tutors bookmark useful websites and consider enabling students to access them
through the school/college intranet.